# Essential Oils - Advances in Extractions and Biological Applications

*Edited by Mozaniel Santana de Oliveira and Eloisa Helena de Aguiar Andrade*

Published in London, United Kingdom

Essential Oils - Advances in Extractions and Biological Applications
http://dx.doi.org/10.5772/intechopen.98130
Edited by Mozaniel Santana de Oliveira and Eloisa Helena de Aguiar Andrade

Contributors
Fatma Guesmi, Ahmed Landoulsi, Hamdy A. Shaaban, Amr Farouk, Ravendra Kumar, Sonu Kumar Mahawer, Himani Himani, Om Prakash, Sushila Arya, Mamdouh S. Salem Serag, Reham A. Elfayoumy, Marwa T. Mohesien, Mejra Bektašević, Olivera Politeo, Sunil Kumar Yadav, Nader Tanideh, Najme Oliyaei, Seyede Zahra Nasirifar, Ishrat Nazir, Sajad Ahmad Gangoo, Nashwa Fathy Sayed Morsy, Sílvia Macedo Arantes, Maria Rosário Martins, Ana Teresa Caldeira, Slimen Selmi, Kais Rtibi, Hichem Sebai, Karim Hosni, Yasvet Yareni Andrade Avila, César Pérez Alonso, Julian Cruz-Olivares



First published in London, United Kingdom, 2022 by IntechOpen
IntechOpen is the global imprint of INTECHOPEN LIMITED, registered in England and Wales, registration number: 11086078, 5 Princes Gate Court, London, SW7 2QJ, United Kingdom
Printed in Croatia

British Library Cataloguing-in-Publication Data
A catalogue record for this book is available from the British Library

Additional hard and PDF copies can be obtained from orders@intechopen.com

Essential Oils - Advances in Extractions and Biological Applications
Edited by Mozaniel Santana de Oliveira and Eloisa Helena de Aguiar Andrade
p. cm.

This title is part of the Biochemistry Book Series, Volume 32
Topic: Chemical Biology
Series Editor: Miroslav Blumenberg
Topic Editors: Şükrü Beydemir and Deniz Ekinci

Print ISBN 978-1-80355-753-3
Online ISBN 978-1-80355-754-0
eBook (PDF) ISBN 978-1-80355-755-7
ISSN 2632-0983

For EU product safety concerns:
IN TECH d.o.o., Prolaz Marije Krucifikse Kozulić 3, 51000 Rijeka, Croatia,
info@intechopen.com or visit our website at intechopen.com.

IntechOpen Book Series

# Biochemistry

Volume 32

## Aims and Scope of the Series

Biochemistry, the study of chemical transformations occurring within living organisms, impacts all of the life sciences, from molecular crystallography and genetics, to ecology, medicine and population biology. Biochemistry studies macromolecules - proteins, nucleic acids, carbohydrates and lipids –their building blocks, structures, functions and interactions. Much of biochemistry is devoted to enzymes, proteins that catalyze chemical reactions, enzyme structures, mechanisms of action and their roles within cells. Biochemistry also studies small signaling molecules, coenzymes, inhibitors, vitamins and hormones, which play roles in the life process. Biochemical experimentation, besides coopting the methods of classical chemistry, e.g., chromatography, adopted new techniques, e.g., X-ray diffraction, electron microscopy, NMR, radioisotopes, and developed sophisticated microbial genetic tools, e.g., auxotroph mutants and their revertants, fermentation, etc. More recently, biochemistry embraced the 'big data' omics systems. Initial biochemical studies have been exclusively analytic: dissecting, purifying and examining individual components of a biological system; in exemplary words of Efraim Racker, (1913 –1991) "Don't waste clean thinking on dirty enzymes." Today, however, biochemistry is becoming more agglomerative and comprehensive, setting out to integrate and describe fully a particular biological system. The 'big data' metabolomics can define the complement of small molecules, e.g., in a soil or biofilm sample; proteomics can distinguish all the proteins comprising e.g., serum; metagenomics can identify all the genes in a complex environment e.g., the bovine rumen.

This Biochemistry Series will address both the current research on biomolecules, and the emerging trends with great promise.

# Meet the Series Editor

Miroslav Blumenberg, Ph.D., was born in Subotica and received his BSc in Belgrade, Yugoslavia. He completed his Ph.D. at MIT in Organic Chemistry; he followed up his Ph.D. with two postdoctoral study periods at Stanford University. Since 1983, he has been a faculty member of the RO Perelman Department of Dermatology, NYU School of Medicine, where he is codirector of a training grant in cutaneous biology. Dr. Blumenberg's research is focused on the epidermis, expression of keratin genes, transcription profiling, keratinocyte differentiation, inflammatory diseases and cancers, and most recently the effects of the microbiome on the skin. He has published more than 100 peer-reviewed research articles and graduated numerous Ph.D. and postdoctoral students.

# Meet the Volume Editors

Mozaniel Santana de Oliveira graduated in Chemistry from the Federal University of Pará, Brazil. He obtained both a master's and Ph.D. in Food Science and Technology from the same university. He has 12 years of professional experience. From 2010 to 2014, he worked on the chemistry of natural products at the Empresa Brasileira de Pesquisa Agropecuária (Embrapa), and from 2014 to 2018, he worked in the Postgraduate Program in Food Science and Technology at the Federal University of Pará, specifically with essential oils. Since 2020, he has been a researcher for the Institutional Training Program - PCI, at the institution Museu Paraense Emilio Goeldi, linked to the Ministério da Ciência, Tecnologia e Inovações of Brazil (MCTI), with studies focused on extraction, characterization chemistry, and applications of essential oils in several industrial segments, among them the food industry. Specifically, Dr. Oliveira has experience in engineering, food science and technology, pharmacology and drug discovery, medicinal chemistry, ethnopharmacology and ethnobotany, phytochemistry, methods of extraction of bioactive compounds, biotechnology of natural products, and allelopathy to find new natural herbicides to control invasive plants. He also has experience in the area of essential oil extraction using supercritical technology and conventional methods. Since 2020, he has supervised and co-supervised master's and Ph.D. students in several graduate programs. Dr. Oliveira serves as a reviewer for thirty-one international scientific journals and is the academic editor of the journals *Evidence-based Complementary and Alternative Medicine*, *Journal of Food Quality*, *Molecules*, and *Open Chemistry*.

Eloisa Helena de Aguiar Andrade holds a degree in Pharmacy (1980), a qualification in Biochemistry (1982), a master's degree in Chemistry of Natural Products (1992), and a Ph.D. in Chemistry (2008) from the Federal University of Pará, Brazil. For. She is currently Associate Researcher II at the Botany Coordination of the Museu Paraense Emílio Goeldi and Adjunct Professor III at the Faculty of Chemistry at the Federal University of Pará. Professor of the Graduate Programs in Chemistry, UFPA, PPG- in Biological C. - Tropical Botany, UFRA/MPEG and Graduate in Biodiversity Biotechnology - Bionorte Network. She is the coordinator of the Pole of the State of Pará, Graduate Program in Biodiversity and Biotechnology (PPG-BIONORTE/PA) of the Bionorte Network (2016-2020). Dr. Andrade is the author of more than 500 scientific contributions, including articles, event communications, book chapters, and books. She has experience in the field of chemistry, with an emphasis on the chemistry of natural products, working mainly on gas chromatography and gas chromatography/mass spectrometry in volatile and fixed chemical constituents (derivatized).

# Contents

# Preface

Humans have always sought out plants as an alternative way to treat disease or for use as food. Plants contain secondary metabolites that have biological activities fundamental for human nutrition and health. In this sense, essential oils have been used for several centuries as agents that promote antibacterial, antifungal, and antiparasitic biological activities. They are also used as insecticides and promoters of phytotoxic activity in medicine and agriculture. Currently, essential oils are widely explored by the pharmaceutical, sanitary, cosmetic, agricultural, and food industries. Traditionally, essential oils are isolated using hydrodistillation. Aromatic plants are the main holders of a wide variety of volatile molecules, being formed mainly of aromatic components derived from phenol or aliphatic molecules.

In addition, advances in research fields have enabled the discovery of numerous chemical compounds produced by plants. This generates concrete perspectives that in the future science will bring new biotechnological processes and applications of these substances to improve quality of life.

*Essential Oils - Advances in Extractions and Biological Applications* includes twelve chapters divided into three sections. The first section talks about the general concepts of essential oils and techniques for their extraction. The second section address topics in food science and technology. The third section discusses essential oils and their pharmacological properties in various activities and applications.

This book provides readers with an understanding of the importance that essential oils have for the safety of plants and the possibilities of extraction and applications.

**Mozaniel Santana de Oliveira**
Coordenação de Botânica,
Museu Paraense Emilio Goeldi,
Belém, Brazil

**Eloisa Helena de Aguiar Andrade**
Federal University of Para,
Belém, Brazil

Section 1

# Essential Oils - Concept and Extraction

Chapter 1

# Extractions Methods and Biological Applications of Essential Oils

*Sonu Kumar Mahawer, Himani, Sushila Arya, Ravendra Kumar and Om Prakash*


## Abstract

Plants produce secondary metabolites for defense and based on the biosynthetic pathway, these chemical compounds are broadly divided into three categories namely nitrogen-containing compounds, phenolic compounds, and terpenes. Essential oils and other such compounds are known for their biological activities. The extraction of essential oils is a challenging aspect for researchers in the field of natural products. Hydrodistillation is a time-consuming and very tedious method. Nowadays, accelerated solvent extraction, supercritical fluid extraction, subcritical water extraction, microwave hydrodiffusion are promising alternatives for conventional methods with several advantages. Essential oils have several biological activities in the field of pharmacological, ethnopharmacological, pesticidal, etc.


**Keywords:** essential oils, secondary metabolites, hydrodistillation, biological activities, accelerated solvent extraction

## 1. Introduction

Essential oils are the highly concentrated, fragrant oil of plant origin that are obtained by steam distillation, dry distillation, hydrodiffusion, or other suitable mechanical methods without heating. These are also denoted as plant "essences" in aromatherapy literature and the method of extraction is critical to categorizing an aromatic constituent as an essential oil [1]. Chemically, these are the mixture of several terpenes or terpenoids which are the polymers of isoprene units. Essential oils are synthesized in the cytoplasm and are usually present in the form of minute droplets between cells. These are insoluble in water, lipophilic and soluble in organic solvents, volatile and aromatic in nature. Almost all plant parts, such as leaves barks, flowers, rhizomes or roots, peels, seeds buds, are reported as the source of essential oils, and several techniques are also known to obtain the essentials from different plant parts [2].

The plant families encompassing species known for most economically significant essential oils are not limited to one taxonomic group only but these are found in all plant classes—gymnosperms such as Pinaceae and Cupressaceae families; angiosperms such as Magnoliopsida, Liliopsida, and Rosopsida. The most important plant families among dicots are Apiaceae, Compositae, Germiniaceae,

Illiciaceae, Lamiaceae, Lauraceae, Myristicaceae, Myrtaceae, Oleaceae, Rosaceae, Santalaceae, etc. whereas, among monocots Zingiberaceae, Poaceae and Acoraceae are the important families [3].

The variations in chemical properties of essential oils vary with their chemical composition and their stereochemical structures, the chemical composition may vary in respect to types of chemical components and their stereochemical nature with the extraction methods used along with the plant type, age, climatic conditions, growth stage, altitude, etc. [4]. Essential oils are the plant secondary metabolites synthesized in the plant cell via metabolic pathways derived from the primary metabolic pathways the synthesis of these metabolites in the plants is often under stress (abiotic and/or biotic) conditions, primarily intervened by different signaling molecules [5] and have been reported for several biological activities which depend upon the chemical composition and stereochemical nature of constituent compounds in essential oils.

In this chapter, we are focusing on the basic information of essential oils, their extraction methods available, and their biological activities in the pharmacological, antimicrobial, and crop protection in agents in agricultural fields.

## 2. Essential oils and their chemical constituents

Essential oils are complex mixtures composed of terpenoids and nonterpenoid volatile hydrocarbons. The basic building unit of essential oils is called an isoprene unit ($C_5H_8$; 2-methyl-1,3-butadiene) and these are arranged following the isoprene rule in a head-to-tail fashion. There are some functional groups also attached which contribute to the biological activities of the essential oils. These groups are mainly alcohols, aldehydes, esters, ethers, ketones, and phenols [6]. Among terpenes, there are subclasses as monoterpenes, sesquiterpenes, diterpenes in the essential oils. Mono terpenes are the results of the combination of two isoprene units, similarly, sesquiterpenes have resulted from three and diterpenes are from four isoprene units. Alcohols, ketones, and carboxylic acids are the functional groups found in the oxygenated derivatives of terpenes, which are jointly known as terpenoids. Apart from terpenes, alcohols, ketones, esters are also found in essential oils as a single component or in combination with terpenes. The basic classification of terpenes is given in **Table 1**.

| S. No. | Terpenes | Number of isoprene unit(s) | Number of carbon atoms | Example(s) |
|---|---|---|---|---|
| 1 | Monoterpenes | 2 | 10 | pinene, myrcene, limonene, thujene |
| 2 | Sesquiterpenes | 3 | 15 | bisabolene, zingiberene, germacrene, caryophyllene |
| 3 | Diterpenes | 4 | 20 | retinol, taxol, and phytol |
| 4 | Triterpenes | 6 | 30 | squalene, hopane |
| 5 | Tetraterpenes | 8 | 40 | carotene, lycopene, and bixin |
| 6 | polyterpenes | >8 | >40 | natural rubber |

**Table 1.**
*Classification of terpenes.*

## 3. Extraction methods

Various parts of aromatic plants can be extracted to obtain the essential oils. Choice of extraction method depends upon the characteristics and components needed for the purposes. In some circumstances, improper and unsuitable extraction techniques can destruct and alter the biological activity of chemical compounds present in essential oils, for example, loss of active compounds, staining, off flavor, and in some cases physical changes in essential oils. For effective extraction with high efficiency, low cost and less tedious methods are required to obtain the high-quality essential oils with high production yield. There are numerous methods that are available for the extraction of essential oils from different parts of plants.

These methods can be grouped into two categories; conventional methods and advanced methods.

### 3.1 Convention extraction methods

#### *3.1.1 Hydrodistillation*

Hydrodistillation is the oldest and most basic oils extraction method which was discovered by Avicenna. The process of extracting essential plant oils by hydrodistillation begins with the plant materials being immersed directly in water inside the vessel and then boiling the entire combination. The devices consist of a heating source, vessel (Alembic), a condenser to convert vapor into liquid, and a decanter to collect the condensate and to separate essential oils with water [7]. This extraction process is a one-of-a-kind way to extract plant materials, such as wood or flowers, and it is commonly employed for extractions requiring hydrophobic natural plant material with a high boiling point. Because the oils are surrounded by water, this process allows essential oils to be extracted at a controlled temperature without overheating. The extraction principle is based on isotropic distillation. Water or other solvents, as well as oil molecules, are present under atmospheric pressure and during the extraction process (heating). The capacity to isolate plant components below 100°C is the fundamental benefit of this extraction approach [8].

#### *3.1.2 Steam distillation*

Steam distillation is a form of distillation or separation technique for temperature-sensitive compounds that are insoluble in water and may break down at their boiling points, such as oils, resins, and hydrocarbons. The basic principle of steam distillation is that it allows a mixture of compounds to be distilled at a temperature that is significantly lower than the individual constituent's boiling point. These compounds, on the other hand, are volatilized at a temperature close to 100°C under atmospheric pressure in the presence of steam or boiling water, by heating plant materials with steam generated by a steam generator. Heat is the primary determinant of how well plant material structures degrade and rupture, releasing aromatic components or essential oils in vapor form [9]. The steam condenses into water when it cools. The film on the water surface (distillate/hydrosol) is then decanted from the top to separate the essential oil from it. At its most basic level, steam distillation has the advantage of being a reasonably inexpensive process to operate, and the qualities of the oils produced by this approach are well known [10]. Masango designed a novel steam distillation extraction technique to enhance separated essential oil yields while reducing wastewater production during the extraction process. A packed sheet of plant samples is put above the steam source in this setup. Boiling water is not allowed to combine with the botanical components, and only steam is allowed to travel through the plants. As a result, less steam is required in the process, and the amount of water in the distillate can be lowered [11]. In another study, by adopting the steam distillation extraction procedure, Yildirim et al. reported a component 2,2- diphenyl-1-picryl hydrazyl (DPPH) that was utilized to evaluate the antioxidant activities of essential oils [12]. It was shown to have a higher yield of antioxidant components than hydrodistillation-extracted oils.

#### *3.1.3 Cold expression*

In the cold expression method, oil is expeller-pressed at low temperatures and pressure. This method ensures that the resultant oil is 100% pure and keeps all of the plant characteristics. It is a mechanical extraction method in which heat is lowered and minimized throughout the raw material batching process. This process is mostly used to extract essential oils from plants, flowers, seeds, and citrus oils, such as lemon and tangerine [13]. In this process, scrubbing is used to remove the outer layer of the plants that contain the oil. The entire plant is then crushed to extract the substance from the pulp and the essential oil from the vesicles. By centrifugation, the essential oil rises to the surface of the substance and is separated from it [14]. The oils derived in this manner have a short shelf life. According to reports, oil produced in this manner contains more of the fruit aromatic character than oil made any other way.

#### *3.1.4 Destructive distillation*

Only birch (*Betula lenta* or *Betula alba*) and cade trees (Juniperus oxycedrus) are used to extract using this approach. After enduring a destructive process under tremendous heat, the hardest components of these woods (e.g., barks, boughs, and roots) are subjected to dry distillation through tar. After condensation, decantation, and separation, a characteristic leathery and empyreumatic oil is formed [15].

#### *3.1.5 Hydrodiffusion*

Contrasting to steam distillation, the steam in this technique is fed from the top to the bottom of the alembic. Through a perforated tray, the vapor mixture, including Eos, is directly condensed underneath the plant support. Separating EOs is done in the same way as previous distillation processes. In comparison to steam distillation, this approach can minimize steam usage and distillation time while also providing a higher yield [15].

### 3.2 Advanced extraction methods

Considering the concepts of economically sound, eco-friendly, high efficiency, and quality production, the efforts were made with respect to the extraction techniques for essential oils from plants.

#### *3.2.1 Microwave-assisted extraction (MAE)*

A microwave is a contactless heat source that can attain more effective and selective heating. Microwaves can complete the distillation in minutes instead of several hours in the conventional distillation method. In this method, plant materials are subjected to a microwave reactor with or without organic solvents or water under different levels of microwave treatments, according to the required protocols [15]. Nowadays, this technique is of high attention among researchers because of its unique heating mechanism (based on friction), cost-effectiveness, high efficacy under normal conditions, higher extraction yield, less extraction times, and high selectivity. Several studies have been reported on the extraction of essential oils using MAE. Recently in 2020, Drinić et al. performed the microwave-assisted hydrodistillation (MAHD) of *Origanum vulgare* L. *spp. hirtum* essential oil and compared with conventional hydrodistillation (HD) using Clevenger-type apparatus [16]. For MAHD, they used an apparatus consisting of a microwave oven connected with a Clevenger-type apparatus. The water to plant ratio was kept similar for both HD and MAHD, 20:1 (w/w). MAHD was accomplished at three different power levels (180, 360, and 600 W) till no more essential oil was obtained. MAHD was found to be a method with several pros over conventional HD. MAHD was found to have less extraction time (24–45 min) as compared to HD

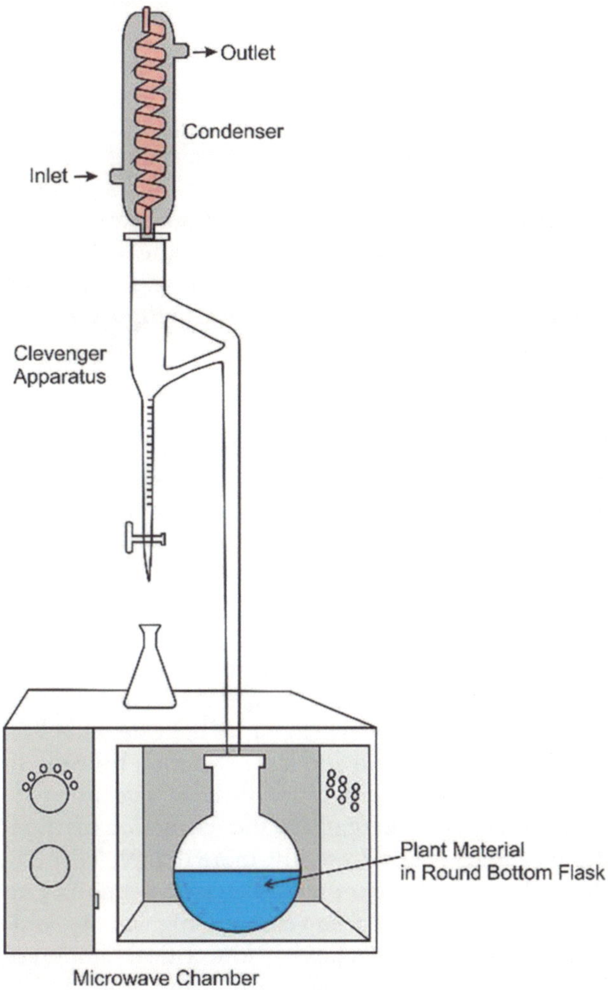


**Figure 1.**
*Microwave-assisted extraction setup.*

(136 min), higher yields of essential oil (2.55–7.10%) as compared to HD (5.81%), higher oxygenated compounds percentage (78.89–85.15%) as compared to (76.82), and it was proven to be a more eco-friendly method (in terms of electrical consumption (0.135–0.240 kW h) as compared to HD (1.360 kW h). Several authors have been published regarding the optimization of extraction procedure using MAE techniques, for instance, successive microwave-assisted extraction optimization for essential oil from lemon peels waste [17] and microwave extraction of essential oils from Eucalyptus globulus leaves [18], and many more. A schematic diagram of the microwave-assisted extraction setup is depicted in **Figure 1**.

### *3.2.2 Ultrasound-assisted extraction*

Similar to MAE, ultrasound-assisted extraction has also been developed to enhance the efficacy along with the reduction in extraction time. The collapse of cavitation bubbles generated through ultrasonication mass transfer and release rate of essential oils get increased by the breakdown of cavitation bubble generated and this cavitation effect is largely depending upon different parameters, such as frequency and intensity of ultrasound, incubation time, temperature in UAE, there are less chances of thermal breakdown, and quality and flavor remain better of essential oils [19, 20].

In a study, ultrasound-assisted hydrodistillation was performed to increase the yield of essential oil from cinnamon bark [21]. They optimized several parameters, and the method developed was compared with conventional hydrodistillation. They found an enhanced yield of essential oils along with a significant reduction in extraction time. Moreover, the scrutiny of electricity utilization and CO2 production demonstrates the eco-friendly and economically soundness of ultrasound-assisted hydrodistillation procedure over conventional hydrodistillation. A schematic diagram of the ultrasound-assisted extraction setup is shown in **Figure 2**.

### *3.2.3 Supercritical fluid extraction (SFE)*

When temperature and pressure are increased over critical points for a given liquid or gas, a supercritical fluid (SF) occurs. The boundary between liquid and gas vanishes in the supercritical zone, and a homogenous fluid arises. Supercritical fluids have a diffusivity and density that distinguishes them from liquids and gases. In contrast to liquids, the density of SFs varies when pressure and temperature values vary, hence a little rise in pressure can result in a massive increase in fluid density, followed by a change in the SF's solvating power. This phenomenon allows for the extraction of specific components from a multicomponent mixture. As a result, supercritical fluid extraction's key benefit is selectivity. The use of this technique may help in the extraction of natural products which have a chance to be degraded at high temperatures. Along with high extraction yield and less extraction time required, this method also allows to recover the solvent used because of the SF's volatile nature, which makes it an economic and environmentally sound extraction method for essential oils and other natural products [22]. Presently, more than 90% of SFE activities are carried out by using $CO_2$, for a variety of uses. $CO_2$ is abstemiously nonflammable, nonexplosive, nontoxic, accessible at cheap cost and high purity, and readily removed from extracts, in addition to having a relatively low critical temperature (32°C) and pressure (7.4 MPa). $CO_2$ also has low surface tension and viscosity but has a diffusivity that is two or three times that of other fluids [23]. $CO_2$ has a polarity similar to pentane in the supercritical zone, making it appropriate for lipophilic compounds

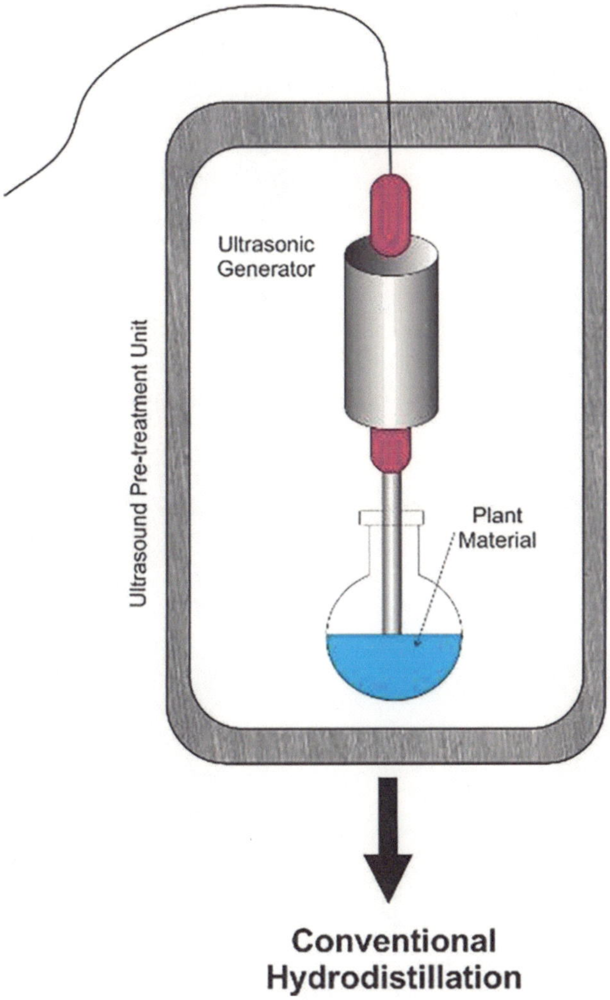


**Figure 2.**
*Schematic diagram of ultrasound-assisted extraction method of essential oils from plant materials.*

extraction. $CO_2$'s fundamental flaw is that it lacks the polarity needed to remove polar compounds [24]. Regarding practical application, in a study, the supercritical $CO_2$ extraction was optimized for the extraction of flower essential oil of the tea (*Camellia sinensis* L.) plants. As per the results showed, the optimum conditions were observed as—pressure of 30 MPa, temperature of 50°C, static time of 10 min, and dynamic time of 90 min for successful extraction of essential oils from the flowers of the tea plant in the sufficient amount [25]. A schematic presentation of the supercritical fluid extraction technique is shown in **Figure 3**.

*3.2.4 Subcritical water extraction*

Subcritical water is defined as water with a temperature above boiling point to a critical point (100–374°C) and a pressure high enough to keep the liquid condition. A phase diagram of water is shown in **Figure 4**. At ambient temperature and pressure, the dielectric constant of water remains highest among all the nonmetallic liquids,

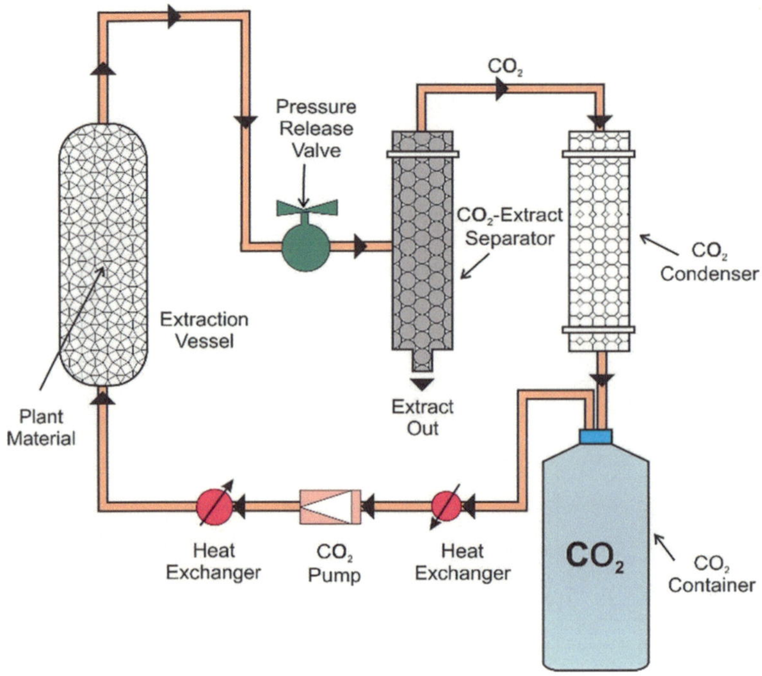


**Figure 3.**
*Schematic diagram of supercritical fluid extraction.*

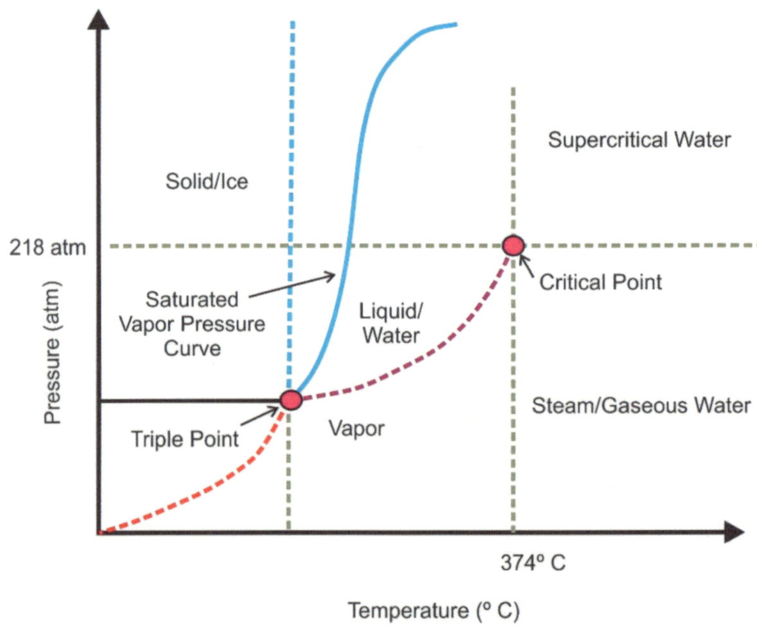


**Figure 4.**
*Phase diagram of water.*

which is reduced significantly in the range of organic solvents, such as acetonitrile, methanol, ethanol, and acetone, and water acts as organic solvent at this temperature and pressure conditions and the plant compounds can be extracted efficiently using

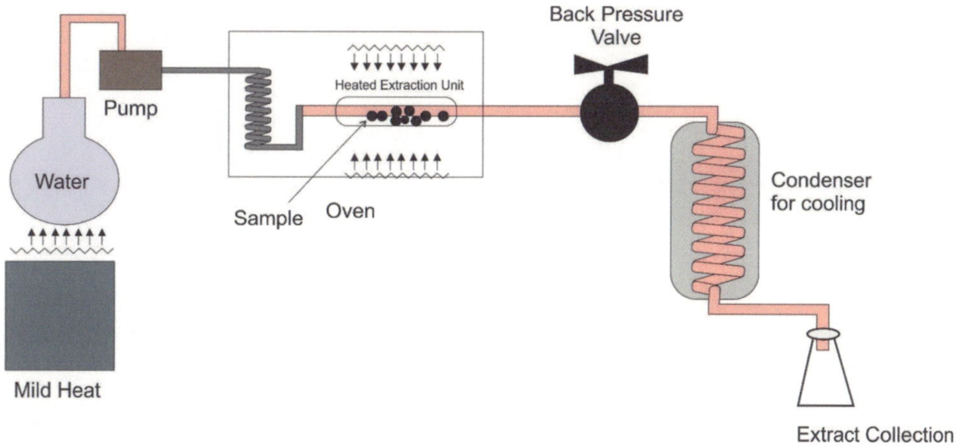


**Figure 5.**
*Simple laboratory setup for SCWE.*

subcritical water (**Figure 5**). The extraction by using subcritical water occurs in the following steps; (1) rapid entry into matrix pores, (2) desorption of solutes from active sites of the matrix, (3) dissolution of solutes in the aqueous fluid, (4) diffusion of solutes through static aqueous fluid in porous materials, (5) diffusion of solutes through the layer of stagnant fluid outside particles and finally, (6) elution of solutes by the flowing of the bulk of aqueous fluid [26]. Currently, SCWE is getting importance in the extraction of essential oils also from different plant parts. For example, SCWE is used to extract essential oils from *Alpinia malaccensis* leaves. The optimum conditions for extraction were found as; the temperature of 156°C, extraction time of 25 minutes. They also reported the interaction of temperature and reaction time parameters using regression analysis. They also conducted kinetics modeling and reported that second-order kinetics was followed by SCWE [27].

#### *3.2.5 Turbo distillation*

The turbo distillation method is similar to conventional water distillation; however, in turbo distillation, the mixture is agitated constantly at a suitable speed using a stainless steel stirrer (**Figure 6**). This approach works well with coarse raw materials and difficult-to-extract substances (spices, woods). When compared to aqueous distillation, turbo distillation reduces distillation durations and energy consumption while also preventing volatile components from degradation. In actuality, it is a type of water distillation-based green extraction [28]. Essential oils can be extracted from difficult-to-extract parts from plants and others using the turbo distillation extraction method. In a study conducted by Mnayer et al., the essential oils and flavonoids were extracted simultaneously using turbo extraction-distillation [29].

#### *3.2.6 Simultaneous distillation extraction*

Likens and Nickerson established simultaneous distillation–extraction (SDE) in 1964, and it has been effectively used to extract essential oils, aromatic compounds, and other volatile products from a variety of matrices. Steam distillation, in combination with continuous extraction with solvent of the mixture of solvents, becomes superior as compared to conventional solvent extraction or extraction with a mixture of solvents.

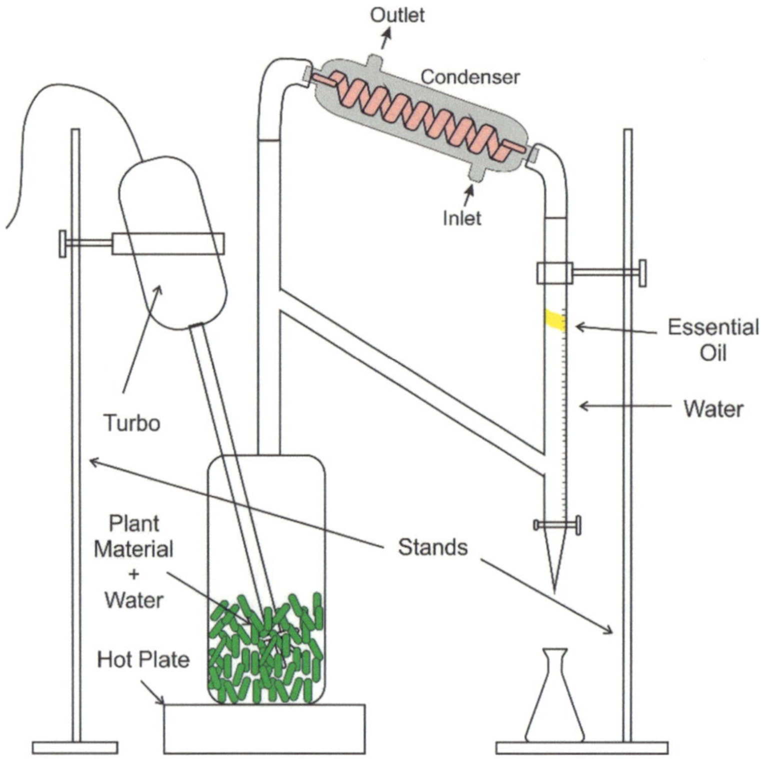


**Figure 6.**
*Turbo hydrodistillation apparatus.*

This is a single-step isolation-concentration technique which reduces extraction time significantly, along with the reduction in the solvents used because of continuous recycling and there is no need for clean up after this approach as the extracts obtained in this technique are devoid of no-volatile components, such as cuticular waxes and chlorophylls [30, 31]. Ribeiro and his coworkers in 2021 performed SDE for the extraction of essential oils from *Rosmarinus officinalis* L [32]. In this study, they assessed the effect of the solvent nature and the optimum time required for extraction. Pentane solvent was found to be best for the performance of SDE for 1 h extraction time.

### *3.2.7 Pulsed electric field-assisted extraction (PEFAE)*

Pulsed electric field (PEF) reduces negative impacts of traditional heating approaches and is a capable substitute to other extraction methods, such as boiling and ultrasound-assisted or microwave-assisted extraction. Moderate to the high strength of the electric field is used on. The PEFAE technique uses moderate to high electric field strength (EFS) ranging from 100 to 300 V/cm and 20 to 80 kV/cm in batch mode and continuous mode extraction, simultaneously. In the electroporation or electropermeabilization (mechanism involved in PEFAE) external electric force is used to augment the cell membrane permeability [33]. The material of interest is kept in between the electrodes and a high-strength electric field in terms of voltage which punctures the cell membrane by the formation of hydrophilic pores and the membrane, its physical functionality and the extraction takes place [34]. PEFAE is of two types viz. batch PEFAE and continuous PEFAE. In Batch PEFAE, the simple is firstly treated with a little solvent between two electrodes and then treated samples removed from the pretreatment unit and stirred at different intensities with the help of a magnetic stirrer to check the solvent evaporation. Apart from the promising results, this process increases operational time because of the low capacity of the system. Therefore, continuous PEFAE is mostly in use at the place of the batch process. In continuous PEFAE, the mixture of solvents is pumped into the treatment chamber by a peristaltic pump at a constant fluid velocity [35]. This technique is getting popularized for the intensification of essential oil extraction [36]. In a study, the PEF was explored for the intensification of essential oil extraction from *Marrubium vulgare*. In this study, PEF pretreatment was done for the purpose of improvement in the permeabilization of the biological membranes. The results revealed a significant enhancement in the extraction rate of essential oils.

## 4. Biological applications of essential oils

### 4.1 Pharmacological applications

#### *4.1.1 Anticancer mechanism of essential oils*

In most cancer chemotherapies, highly cytotoxic drugs are used that target proliferating cell populations. The nondiscriminatory nature of these drugs leads to severe side effects in normal cells. Natural essential oils and their constituents play a significant role in cancer prevention and treatment. Various mechanisms are responsible for the chemopreventive properties of essential oils, such as antioxidant, antiproliferative, and antimutagenic, enhancing detoxification and synergistic action of their constituents. There is a direct relationship between the production of reactive oxygen species to the origin of oxidation and inflammation that can lead to cancer. Mitochondrial DNA damage can result from oxidative stress which leads to an increase in the mutation rate within cells and thus promoting oncogenic transformation. Besides this, reactive oxygen species (ROs) specifically activate signaling pathways and promote tumor development through the regulation of cellular proliferation, angiogenesis, and metastasis [37]. EOs components react with ROs and form reactive phenoxy radicals which can then react with further ROs to prevent further [38]. EOs also induces the expression of antioxidant enzymes, such as catalase, superoxide dismutase, glutathione peroxidase, and glutathione, which leads to an increase in intracellular antioxidant activity, subsequently leading to a significant reduction in tumor mass (**Figure** 7). Several studies have shown the anticancer activity of EOs against various cancers, some are summarized in **Table 2**.

#### *4.1.2 Essential oil as an antioxidant agent*

Free radicals and other reactive oxygen species cause oxidation of biomolecules which ultimately leads to molecular alterations, including chronic disorders associated with the aging process, arteriosclerosis and cancer [49], Alzheimer's disease [50], Parkinson's disease, diabetes, and asthma. Essential oils also exhibit remarkable antioxidant activity/free radical scavenging activity which has often been confirmed by physicochemical methods (**Table 3**). The essential oils of some medicinally important plants, such as basil, cinnamon, clove, nutmeg, oregano, and thyme, have proven radical-scavenging and antioxidant properties [63]. The antioxidant properties are mainly dependent on the chemical constituents, such as in *Thymus* the antioxidant activity is mainly attributed to the presence of thymol and carvacrol content (36.5 and 29.8%).

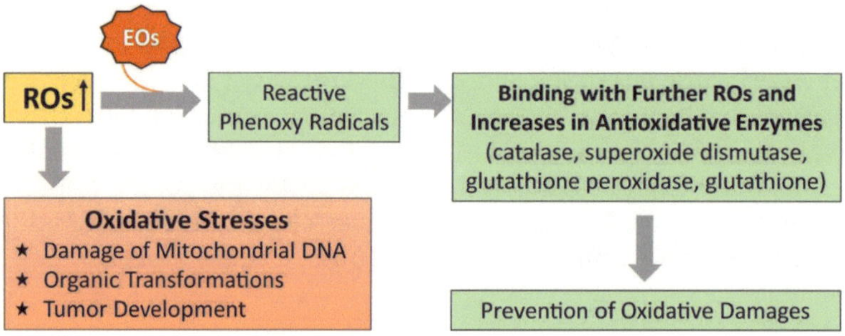


**Figure 7.**
*Antioxidant mechanism responsible for chemo preventive mechanism.*

| Cancer | Species | Major essential oil constituents | Cancer cell lines used | Reference |
|---|---|---|---|---|
| Lung | *Morinda citrifolia* | L-scopoletin, nordamnacanthal, β-morindone, α-copaene, 9-H-pyrido[3,4-*b*]indole, β-thujene | A549 | [39] |
| Breast | *Boswelia sacra* | α-pinene, α-thujene, myrcene, boswelic acid | T47D, MCF7, MDA-MB-231 | [40] |
| Colon | *Citrus limettioides* | d-limonene, triacontane, sabinene, β-myrcene | SW480 | [41] |
| Ovary | *Nepeta ucrainica* | germacrene D, bicyclogermacrene, β-bourbonene, β-elemene, spathulenol, cubenol | A2780 | [42] |
| Liver | *Thymus citriodorus* | borneol, thymol, 3, 7-dimethyl-1, 6-octadiene-3-ol, 1-methyl-4-[alpha-hydroxy-isopropyl] cyclohexene, camphor | HepG2 | [43] |
| Uterus and Cervix | *Casearia sylvestris* | α-zingiberene, *E*-caryophyllene,α-acoradiene, α-muurolol, viridiflorene | Siha | [44] |
| Oral | *Solanium spirale Roxb.* | *E*-phytol, n-hexadecanoic acid, β-selinene, α-selinene | KB | [45] |
| Pancreas | *Angelica archangelica* | β-phellandrene, α-pinene, bicyclogermacrene, sabinene, bicycloelemene, | PANC-1 | [46] |
| Leukemia | *Malus domestica* | eucalyptol, phytol, α-farnesene, pentacosane | THP-1 | [47] |
| Skin | *Schefflera heptaphylla* | β-pinene, 4-methyl-1-(1-methylethyl)-3-cyclohexene, 3,7-dimethyl-1,6-octadien, β-caryophyllene | A375 | [48] |

**Table 2.**
*Anticancer activity of EOs against various cancer cell lines.*

| Plant | Family | Major essential oil constituents | Antioxidant assay | Reference |
|---|---|---|---|---|
| *Limnophilla indica* | Plantaginaceae | *epi*-cyclocolorenone, α-gurjunene, 5-hydroxy-cis-calamenene, α-gurjunene, β-caryophyllene, | DPPH radical scavenging, Reducing power, Metal chelating of $Fe^{2+}$ | [51] |
| *Zanthoxylum armatum* | Rutaceae | α-pinene, germacrene-D, *E*-caryophyllene, α-cadinol, 2-undecanone | DPPH radical scavenging, Reducing power, Metal chelating of $Fe^{2+}$ | [52] |
| *Premna mucronata* | Lamiaceae | ethyl hexanol, 1-octen-3-ol, linalool, methyl salicylate, *E*-caryophyllene | DPPH radical scavenging, Reducing power, Metal chelating of $Fe^{2+}$ | [53] |
| *Salvia reflexa* | Lamiaceae | palmitic acid, phytol, *E*-caryophyllene caryophyllene oxide | DPPH radical scavenging, Reducing power, Metal chelating of $Fe^{2+}$ | [54] |
| *Coleus barbatus* | Lamiaceae | bornyl acetate, n-decanal, sesquisabinene, β-bisabolene, δ-cadinene | DPPH radical scavenging, Reducing power, Metal chelating of $Fe^{2+}$, $H_2O_2$ radical scavenging, Nitric oxide radical scavenging, Superoxide radical scavenging | [55] |
| *Globba sessiliflora* | Zingiberaceae | myrcene, β-caryophyllene, selin-11-en-4α-ol, β-longipinene, manool, germacrene D and β-eudesmol | DPPH radical scavenging, Reducing power, Metal chelating of $Fe^{2+}$ | [56] |
| *Caryopteris foetida* | verbenaceae | δ-cadinene, farnesene, β-caryophyllene, γ-cadinene,spathulenol, τ-muurolol | DPPH radical scavenging, Reducing power, Metal chelating of $Fe^{2+}$ | [57] |
| *Nepeta cataria* | Lamiaceae | cis-nepetalactone, bicyclo [3.1.0] hexane-2-undecanoic acid, methyl ester, (*cis-,trans*) nepetalactone, | DPPH radical scavenging, Reducing power, Metal chelating of $Fe^{2+}$ | [58] |
| *Cotinus coggygria* | Anacardiacea | myrcene, α-pinene, α-terpineol, cymene, sabinene | DPPH radical scavenging, Metal chelating of $Fe^{2+}$ | [59] |
| *Foeniculum vulgare* | Apiaceae | limonene, estragole, *trans*-anethole, fenchone, eucalyptol | DPPH free radical scavenging, ferric reducing power assay, thiobarbituric acid reactive species assay, ferrous ion-chelating | [60] |
| *Origanum vulgare* | Lamiaceae | carvacrol, thymol, p-cymene, caryophyllene, 3-carene | DPPH radical scavenging, Reducing power | [61] |

| Plant | Family | Major essential oil constituents | Antioxidant assay | Reference |
|---|---|---|---|---|
| *Ocimum basilicum* | Lamiaceae | linalool, methyl chavicol, eucalyptol, eugenol, *trans*-α-bergamotene | DPPH radical scavenging, β-carotene bleaching assay, TBHQ inhibition | [62] |

**Table 3.**
*Reported antioxidant activities of essential oils of different plant families.*

### 4.1.3 Essential oil as an antidiabetic agent

Hyperglycemia is a condition of diabetes that arises as a result of the inability to either produce insulin or use it to regulate normal glucose levels in the blood. Inhibition of α-glucosidase and α-amylase is an important factor to control postprandial hyperglycemia in the management of type 2 diabetes mellitus as both enzymes are involved in the digestion of carbohydrates. α-amylase is involved in the break down of long-chain carbohydrates into disaccharides while α-glucosidase breaks down starch and disaccharides to glucose or monosaccharides. Thus, by inhibiting the enzyme, carbohydrate breakdown can be delayed, and ultimately absorption of glucose in the bloodstream is reduced [64]. Essential oils bind to the active site of the enzyme (α-amylase or α-glucosidase) and act as an inhibitor to form an enzyme-inhibitor complex thus inhibiting the enzyme activities (**Figure 8**).

Several essential oils and their constituents have been analyzed for their antidiabetic potential such as essential of plant *Syzygium aromaticum, Cuminum cyminum* [66], *Nepeta hindostana* [57], *Oliveria decumbens*, *Thymus kotschyanus*, *Trachyspermum ammi*, *Zataria* multiflora [67], and *Carthamus tinctorius* [68].

## 4.2 Antimicrobial application

EOs and their constituents play a vital role in possessing antimicrobial activities. The antimicrobial activity of essential oil mainly depends on three characteristics—the nature of EO (hydrophilic or hydrophobic), its chemical constituents, and the targeted organism [69, 70]. Due to their hydrophobic nature, EOs passes across the cell wall and cytoplasmic membrane and disrupt the cell wall structure and make them more permeable. The membrane permeability leads to leakage of macromolecules and

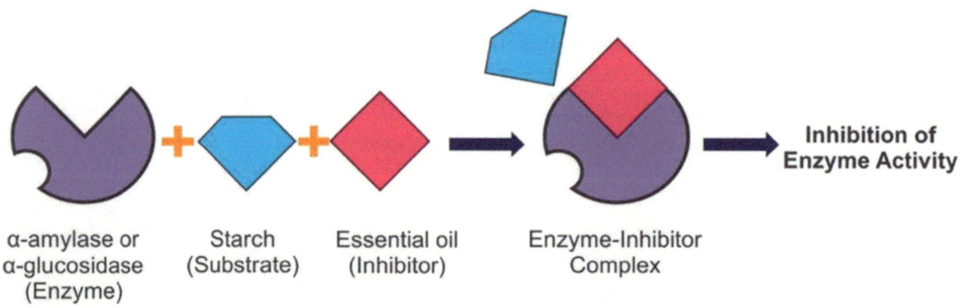


**Figure 8.**
*Mechanism of enzyme (α-amylase and α-glucosidase) inhibition by essential oil [65].*

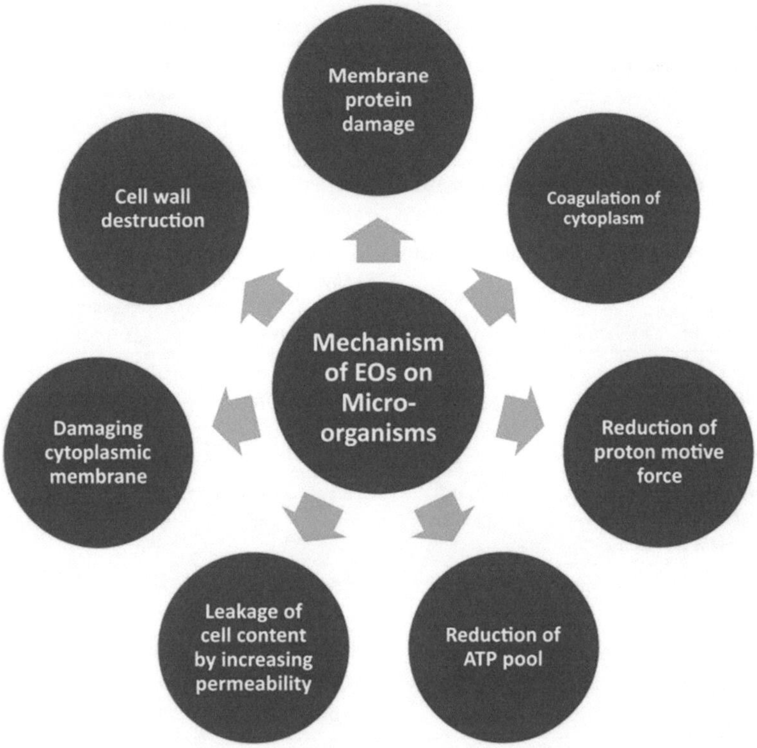


**Figure 9.**
*Mechanism of essential oil action on micro-organisms.*

other cellular materials leading to cell death [71]. In bacteria, the permeabilization of the membranes is associated with loss of ions and reduction of membrane potential, the collapse of the proton pump, and depletion of the ATP pool. EOs can also damage lipids and proteins by coagulating the cytoplasm. **Figure 9** represents different kinds of the mechanism of action of essential oils on microorganisms. The antimicrobial activity of EO's is mostly due to the presence of phenols, aldehydes, and alcohols. Terpenoids are one of the major constituents present in EOs and have oxygen atoms or methyl groups, which are localized or removed from specific enzymes by which they show greater activities. Generally, it has been observed that EOs are more active in gram-positive bacteria than gram-negative bacteria due to the presence of peptidoglycan layer which lies outside the outer membrane. Whereas, in gram-negative bacteria, the outer membrane is composed of a double layer of phospholipids and it is linked with the inner membrane by lipopolysaccharides thus hydrophobic macromolecules, such as essential oils constituents are unable to penetrate the membrane which is responsible for the resistance of the gram-negative bacteria to EOs. Aflatoxins, which are toxic secondary metabolites produced by common fungi, such as *Aspergillus flavus* and *A. parasiticus,* cause contamination of many food products. These aflatoxins are teratogenic, carcinogenic, and mutagenic. Some essential oils not only inhibit the growth of such fungi but can also stop the production of aflatoxins. EOs are effective against a wide range of plants and human pathogenic bacteria, fungi, and viruses by using different assays, as summarized in **Table 4**.

| Essential oil | Action | Target microorganism | Reference |
|---|---|---|---|
| *Cinnamomum osmophloeum* | Antibacterial | *Escherichia coli, Pseudomonas aeruginosa, Enterococcus faecalis, Staphylococcus aureus, Klebsiella pneumoniae, Salmonella sp, Vibrio parahemolyticus* | [72] |
| *Mentha piperita, Ocimum basilicum* L | Antibacterial | *E. coli O157:H7 and Salmonella typhimurium* | [73] |
| *Thymus vulgaris, Pimpinella anisum seeds* | Antibacterial | *S. aureus, Bacillus cereus, E. coli, Proteus vulgaris, Proteus mirabilis, Salmonella typhi, S. typhimurium, K. pneumoniae and P. aeruginosa* | [74] |
| *Allium sativum, Artemisia Judaica, Satureja hortensis, Rosmarinus officinalis Cedrus libani, Chenopodium ambrosioides* | Antifungal | *Aspergillus niger, Aspergillus parasiticus, Aspergillus flavus, Aspergillus fumigatus,* | [75] |
| *R. officinalis, Foeniculum vulgare, Artemisia Judaica, Artemisia absinthium, Artemisia biennis* | Antifungal | *Botrytis cinerea; Botrytis fabae* | [75] |
| *Artemisia judaica, A. absinthium, A. biennis* | Antifungal | *Pythium debaryanum, Trichophyton rubrum; T. mentagrophytes; T. roseum* | [76] |
| *A. sativum, Artemisia judaica, A. absinthium, A. biennis* | Antifungal | *Penicillium cyclopium; Fusarium oxysporum;* | [76–78] |
| *Citrus essential oil* | Antiviral | SARS-CoV2 | [79] |
| *Illicium verum* | Antiviral | Herpes simplex virus | [80] |
| *Allium cepa, A. sativum, Cuminum cyminum, Corriandrum sativum, Petroselinum sativum, O. basilicum* | Antiviral | Herpes simplex virus | [81] |

**Table 4.**
*Antimicrobial activities of essential oils.*

### 4.3 Pesticidal applications of essential oils

Pesticides include a wide range of compounds with very different actions (such as herbicides, insecticides, nematicides, rodenticides, avicides, algicides, fungicides, bactericides, and others) [82]. Due to the high toxicity, environmental pollution, high cost, and many more disadvantages of chemical pesticides, researches are intended toward finding novel solutions with lower toxicity, fewer damaging behavior toward the environment, and a better specificity of action. In this regard, a number of botanicals have historically been used for the control of storage pests, particularly in the Mediterranean region and Southern Asia; however, the importance of essential oils arose in the 1990s following the discovery of their fumigant and contact insecticidal activities against a wide range of pests [83].

Essential oil plays a significant role in the plant's defense against bacteria, viruses, insects, fungi, and herbs. Essential oils are a complex and distinctive mixture of compounds that can be considered for next-generation pesticides. In the case of insecticidal actions, some oils appear to interact with the neuromodulator octopamine, while others appear to interfere with GABA-gated chloride channels, indicating that they have a neurotoxic mechanism of action. With the evidence of the potential of essential oils in pest control, these are considered as new approaches in pest control

| S. No. | Botanical source of essential oil (species) | Pesticidal Action as | Activity against | Reference(s) |
|---|---|---|---|---|
| 1. | *Limnophila indica, Cotinus coggygria* and *Hedychium spicatum* | Antifeedant | larvae of *Spilosoma obliqua* | [51, 59, 85] |
| 2. | *Coleus barbatus* | Antifeedant | third instar larvae of *Spilosoma oblique* | [55] |
| 3. | *Cinnamomum camphora* | Insecticidal and insect repellent | *Aphis gossypii* Glover | [86] |
| 4. | *L. indica, C. coggygria* and *Hedychium spicatum* | Herbicide | *Raphanus raphanistrum* sub sp. *sativus* | [59, 85, 87] |
| 5. | *C. barbatus* | Herbicide | *R. raphanistrum* | [55] |
| 6. | *Cinnamomum zeylanicum* | Acaricide | *Dermatophagoides* spp. and *Tyrophagus putrescentiae* mites | [88] |
| 7. | *Ocimum* L. | Fungicide | *Rhizoctonia solani* | [89] |
| 8. | *Alpinia allughas* and *Alpinia malaccensis* | Fungicide | *Colletotricum falcatum*, *Rhizoctonia solani*, *Sclerotinia sclerotium* and *Sclerotium rolfsii* | [90, 91] |

**Table 5.**
*Pesticidal applications of essential oils from different plant species.*

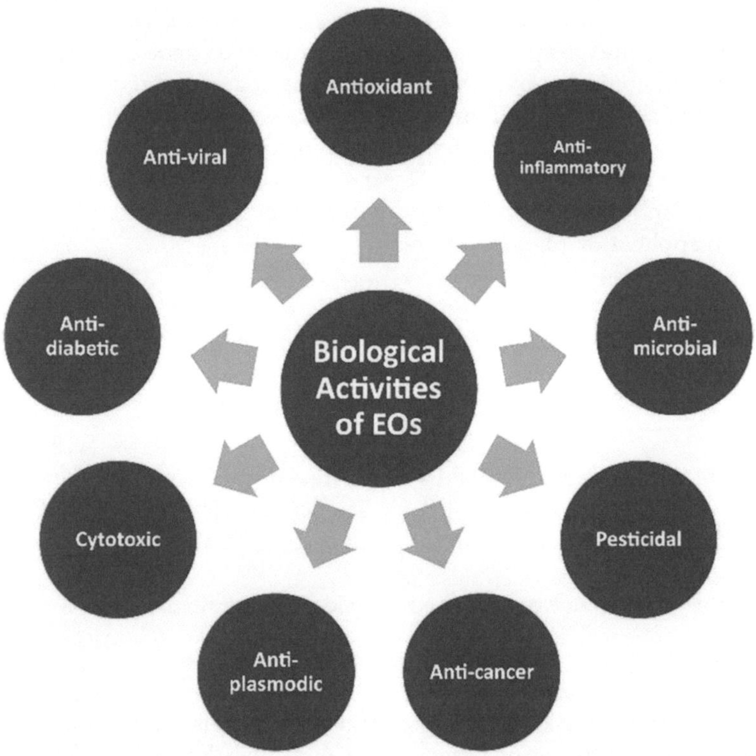


**Figure 10.**
*Various biological activities of essential oils.*

viz. essential oil-based pesticides [84]. There are several applications of essential oils in plant protection, such as insecticidal, herbicidal, nematicidal, and fungicidal. Some such recent studies have been enlisted in **Table 5**.

### 4.4 Other biological activities of EOs

Several pharmaceutical and biological activities, such as antibacterial, antifungal, anticancer, antiviral, antidiabetic, antimutagenic, antiprotozoal, anti-inflammatory, antipyretic, analgesic, hepatoprotective, antidiarrheal, antihyperlipidemic, diuretic, neuroprotective, and pesticidal activities, have been reported in various medicinal and aromatic plants bearing essential oil (**Figure 10**) [75]. Essential oils of different plant families of angiospermic plants possess various therapeutic qualities like medicinal and antimicrobial properties (constipation, dysentery, malaria, measles, stomach pain, yellow fever, and dental care).

## 5. Conclusion

Essential oils are the important secondary metabolites of plants and are found in almost all parts of the plants. There are several extraction techniques are available. Conventional methods, such as hydrodistillation, steam distillation, cold expression, have several disadvantages. To overcome such cons, researchers have been developed several advanced extraction techniques for essential oils viz. microwave-assisted extraction, supercritical fluid extraction, ultrasound-assisted extraction, subcritical water extraction, etc. These advanced methods also have some cons and there is a need to research for the cheap, easy, eco-friendly methods to be developed. In this chapter, various biological applications, such as pharmacological, antimicrobial, pesticidal activities, have been discussed. There are several aromatic plant species that are remained unexploited for their essential oils and their potential in biological application. There is a need to explore newer species of aromatic plants in this regard and further research is needed in the future in respect to the extraction techniques and biological application of essential oils.

## Author details

Sonu Kumar Mahawer, Himani, Sushila Arya, Ravendra Kumar* and Om Prakash
Department of Chemistry, College of Basic Sciences and Humanities, G.B. Pant University of Agriculture and Technology, Pantnagar, Uttarakhand, India

*Address all correspondence to: ravichemistry.kumar@gmail.com

## References

[1] Manion CR, Widder RM. Essentials of essential oils. American Journal of Health-System Pharmacy. 2017;**74**(9): 153-162

[2] Tongnuanchan P, Benjakul S. Essential oils: Extraction, bioactivities, and their uses for food preservation. Journal of food science. 2014;**79**(7):1231-1249

[3] Franz C, Novak J. Sources of essential oils. In: Handbook of Essential Oils. Vol. 10. London: CRC Press; 2020. pp. 41-83

[4] Angioni A, Barra A, Coroneo V, Dessi S, Cabras P. Chemical composition, seasonal variability, and antifungal activity of Lavandula stoechas L. ssp. stoechas essential oils from stem/leaves and flowers. Journal of agricultural and food chemistry. 2006;**54**(12):4364-4370

[5] Hussein RA, El-Anssary AA. Plants secondary metabolites: The key drivers of the pharmacological actions of medicinal plants. Herbal Medicine. 2019;**1**:13

[6] Buckle J. Basic plant taxonomy, basic essential oil chemistry, extraction, biosynthesis, and analysis. Clinical Aromatherapy. 2015. pp. 37-72

[7] Asbahani A, Miladi K, Badri W, Sala M, Addi EHA, Casabianca H, et al. Essential oils: From extraction to encapsulation. International Journal of Pharmaceutics;**483**:220-243

[8] Golmakani MT, Rezaei K. Comparison of microwave-assisted hydrodistillation with the traditional hydrodistillation method in the extraction of essential oils from Thymus vulgaris L. Food Chemistry. 2008;**109**:925-930

[9] Babu KGD, Kaul VK. Variation in essential oil composition of rose scented geranium (Pelargonium sp.) distilled by different distillation techniques. Flavour and Fragrance Journal. 2005;**20**:222-231

[10] Karl-Georg F, Franz-Josef H, Johannes P, Wilhelm P, Dietmar S, Kurt B, et al. Flavors and fragrances. Ullmann's Encyclopedia of Industrial Chemistry. 2003. DOI: 10.1002/14356007.a11_141

[11] Masango P. Cleaner production of essential oils by steam distillation. Journal of Cleaner Production. 2005;**13**:833-839

[12] Yildirim A, Cakir A, Mavi A, Yalcin M, Fauler G, Taskesenligil Y. The variation of antioxidant activities and chemical composition of essential oils of Teucriumorientale L. var. orientale during harvesting stages. Flavour and Fragrance Journal. 2004;**19**:367-372

[13] Arnould-Taylor WE. Aromatherapy for the Whole Person. UK: Stanley Thornes; 1981. pp. 22-26

[14] Rassem HH, Nour AH, Yunus RM. Techniques for extraction of essential oils from plants: A review. Australian Journal of Basic and Applied Sciences. 2016;**10**(16):117-127

[15] Li Y, Fabiano-Tixier AS, Chemat F. Essential oils: From conventional to green extraction. Essential oils as reagents in Green Chemistry. 2014. pp. 9-20

[16] Drinić Z, Pljevljakušić D, Živković J, Bigović D, Šavikin K. Microwave-assisted extraction of O. vulgare L. spp. hirtum essential oil: Comparison with conventional hydro-distillation. Food and Bioproducts Processing. 2020;**120**:158-165

[17] Martínez-Abad A, Ramos M, Hamzaoui M, Kohnen S, Jiménez A, Garrigós MC. Optimisation of sequential microwave-assisted extraction of essential oil and pigment from lemon peels waste. Food. 2020;**9**(10):1493

[18] Tran TH, Ngo TC, Dao TP, Nguyen PT, Pham TN, Nguyen TD, et al. Optimizatoin of Microwave-assisted extraction and compositional determination of essential oil from leaves of Eucalyptus globulus. IOP Conference Series: Materials Science and Engineering. 2020;**736**(2):022040

[19] Asfaw N, Licence P, Novitskii AA, Poliakoff M. Green chemistry in Ethiopia: The cleaner extraction of essential oils from Artemisia afra: A comparison of clean technology with conventional methodology. Green Chemistry. 2005;7(5):352-356

[20] Da Porto C, Decorti D, Kikic I. Flavour compounds of *Lavandula angustifolia* L. to use in food manufacturing: Comparison of three different extraction methods. Food Chemistry. 2009;**112**(4):1072-1078

[21] Chen G, Sun F, Wang S, Wang W, Dong J, Gao F. Enhanced extraction of essential oil from Cinnamomum cassia bark by ultrasound assisted hydrodistillation. Chinese Journal of Chemical Engineering. 2021;**36**:38-46

[22] Yousefi M, Rahimi-Nasrabadi M, Pourmortazavi SM, Wysokowski M, Jesionowski T, Ehrlich H, et al. Supercritical fluid extraction of essential oils. TrAC Trends in Analytical Chemistry. 2019;**118**:182-193

[23] Yamini Y, Khajeh M, Ghasemi E, Mirza M, Javidnia K. Comparison of essential oil compositions of *Salvia mirzayanii* obtained by supercritical carbon dioxide extraction and hydrodistillation methods. Food Chemistry. 2008;**108**(1):341-346

[24] Wang SM, Ling YC, Giang YS. Forensic applications of supercritical fluid extraction and chromatography. Journal of Forensic Sciences. 2003;**2**(5):5-18

[25] Chen Z, Mei X, Jin Y, Kim EH, Yang Z, Tu Y. Optimisation of supercritical carbon dioxide extraction of essential oil of flowers of tea (*Camellia sinensis* L.) plants and its antioxidative activity. Journal of the Science of Food and Agriculture. 2014;**94**(2):316-321

[26] Gbashi S, Adebo OA, Piater L, Madala NE, Njobeh PB. Subcritical water extraction of biological materials. Separation & Purification Reviews. 2017;**46**(1):21-34

[27] Samadi M, Zainal Abidin Z, Yoshida H, Yunus R, Awang Biak DR, Lee CH, et al. Subcritical water extraction of essential oil from *Aquilaria malaccensis* leaves. Separation Science and Technology. 2020;**55**(15):2779-2798

[28] Ghasemy-piranloo F, Kavousi F, Dadashian S. Comparison for the production of essential oil by conventional, novel and biotechnology methods. Authorea Preprints. 2020;**20**:1-34

[29] Mnayer D, Fabiano-Tixier AS, Petitcolas E, Ruiz K, Hamieh T, Chemat F. Simultaneous extraction of essential oils and flavonoids from onions using turbo extraction-distillation. Food Analytical Methods. 2015;**8**(3):586-595

[30] Chaintreau A. Simultaneous distillation–extraction: From birth to maturity. Flavour and Fragrance Journal. 2001;**16**(2):136-148

[31] Teixeira S, Mendes A, Alves A, Santos L. Simultaneous distillation–

extraction of high-value volatile compounds from Cistus ladanifer L. Analytica Chimica Acta. 2007;**584**: 439-446

[32] Ribeiro BS, Ferreira MD, Moreira JL, Santos L. Simultaneous distillation–Extraction of essential oils from rosmarinus officinalis L. Cosmetics. 2021;**8**(4):117

[33] Panja P. Green extraction methods of food polyphenols from vegetable materials. Current Opinion in Food Science. 2018;**23**:173-182

[34] Redondo D, Venturini ME, Luengo E, Raso J, Arias E. Pulsed electric fields as a green technology for the extraction of bioactive compounds from thinned peach by-products. Innovative Food Science & Emerging Technologies. 2018;**45**:335-343

[35] Ranjha MM, Kanwal R, Shafique B, Arshad RN, Irfan S, Kieliszek M, et al. A critical review on pulsed electric field: A novel technology for the extraction of phytoconstituents. Molecules. 2021;**26**(16):4893

[36] Miloudi K, Hamimed A, Benmimoun Y, Bellebna Y, Taibi A, Tilmatine A. Intensification of essential oil extraction of the Marrubium vulgare using pulsed electric field. Journal of Essential Oil Bearing Plants. 2018;**21**(3):811-824

[37] Storz P. Reactive oxygen species in tumor progression. Frontiers in Bioscience. 2005;**10**(1-3):1881-1896

[38] Arenas DRM, Acevedo AM, Méndez LYV, Kouznetsov VV. Scavenger activity evaluation of the clove bud essential oil (*Eugenia caryophyllus*) and eugenol derivatives employing ABTS+• decolorization. Scientia Pharmaceutica. 2011;**79**(4):779

[39] Rajivgandhi G, Saravanan K, Ramachandran G, Li JL, Yin L, Quero F, et al. Enhanced anti-cancer activity of chitosan loaded *Morinda citrifolia* essential oil against A549 human lung cancer cells. International Journal of Biological Macromolecules. 2020;**164**:4010-4021

[40] Suhail MM, Wu W, Cao A, Mondalek FG, Fung KM, Shih PT, et al. *Boswellia sacra* essential oil induces tumor cell-specific apoptosis and suppresses tumor aggressiveness in cultured human breast cancer cells. BMC Complementary and Alternative Medicine. 2011;**11**(1):1-14

[41] Jayaprakasha GK, Murthy KC, Uckoo RM, Patil BS. Chemical composition of volatile oil from *Citrus limettioides* and their inhibition of colon cancer cell proliferation. Industrial Crops and Products. 2013;**45**:200-207

[42] Shakeri A, Khakdan F, Soheili V, Sahebkar A, Rassam G, Asili J. Chemical composition, antibacterial activity, and cytotoxicity of essential oil from *Nepeta ucrainica* L. spp. *kopetdaghensis*. Industrial Crops and Products. 2014;**58**:315-321

[43] Wu S, Wei FX, Li HZ, Liu XG, Zhang JH, Liu JX. Chemical composition of essential oil from *Thymus citriodorus* and its toxic effect on liver cancer cells. Journal of Chinese Medicinal Materials. 2013;**36**(5):756-759

[44] Bou DD, Lago JHG, Figueiredo CR, Matsuo AL, Guadagnin RC, Soares MG, et al. Chemical composition and cytotoxicity evaluation of essential oil from leaves of *Casearia sylvestris*, its main compound α-zingiberene and derivatives. Molecules. 2013;**18**(8):9477-9487

[45] Keawsa-Ard S, Liawruangrath B, Liawruangrath S, Teerawutgulrag A, Pyne SG. Chemical constituents and antioxidant and biological

activities of the essential oil from leaves of *Solanum spirale*. Natural Product Communications. 2012;**7**(7):1934578X1200700740

[46] Sigurdsson S, Ogmundsdóttir HM, Gudbjarnason S. The cytotoxic effect of two chemotypes of essential oils from the fruits of *Angelica archangelica* L. Anticancer Research. 2005;**25**(3B):1877-1880

[47] Walia M, Mann TS, Kumar D, Agnihotri VK, Singh B. Chemical composition and in vitro cytotoxic activity of essential oil of leaves of *Malus domestica* growing in Western Himalaya (India). Evidence-Based Complementary and Alternative Medicine. 2012;**2012**:1-6

[48] Li YL, Yeung CM, Chiu LC, Cen YZ, Ooi VE. Chemical composition and antiproliferative activity of essential oil from the leaves of a medicinal herb, *Schefflera heptaphylla*. Phytotherapy Research: An International Journal Devoted to Pharmacological and Toxicological Evaluation of Natural Product Derivatives. 2009;**23**(1):140-142

[49] Gardner P. Superoxide-driven aconitase FE-S center cycling. Bioscience Reports. 1997;**17**:33-42

[50] Butterfield D, Lauderback C. Lipid peroxidation and protein oxidation in Alzheimer's disease brain: Potential causes and consequences involving amyloid beta-peptideassociated free radical oxidative stress. Free Radical Biology & Medicine. 2020;**32**:1050-1060

[51] Kumar R, Kumar R, Prakash O, Srivastava RM, Pant AK. Chemical composition, in vitro antioxidant, anti-inflammatory and antifeedant properties in the essential oil of Asian marsh weed *Limnophila indica* L.(Druce). Journal of Pharmacognosy and Phytochemistry. 2019;**8**(1):1689-1694

[52] Dhami A, Singh A, Palariya D, Kumar R, Prakash O, Rawat DS, et al. α-pinene rich bark essential oils of *Zanthoxylum armatum* DC. from three different altitudes of Uttarakhand, India and their antioxidant, in vitro anti-inflammatory and antibacterial activity. Journal of Essential Oil Bearing Plants. 2019;**22**(3):660-674

[53] Palariya D, Singh A, Dhami A, Pant AK, Kumar R, Prakash O. Phytochemical analysis and screening of antioxidant, antibacterial and anti-inflammatory activity of essential oil of *Premna mucronat*a Roxb. leaves. Trends in Phytochemical Research. 2019;**3**(4):275-286

[54] Goswami S, Kanyal J, Prakash O, Kumar R, Rawat DS, Srivastava RM, et al. Chemical composition, antioxidant, antifungal and antifeedant activity of the *Salvia reflexa* hornem. Essential oil. Asian Journal of Applied Sciences. 2019;**12**(4):185-191

[55] Kanyal J, Prakash O, Kumar R, Rawat DS, Srivastava RM, Singh RP, et al. Study on comparative chemical composition and biological activities in the essential oils from different parts of *Coleus barbatus* (Andrews) Bent. ex G. Don. Journal of Essential Oil Bearing Plants. 2021;**24**(4):808-825

[56] Kumar R, Prakash O, Pant AK, Isidorov VA, Mathela CS. Chemical composition, antioxidant and myorelaxant activity of essential oils of *Globba sessiliflora* Sims. Journal of Essential Oil Research. 2012; **24**(4):385-391

[57] Joshi A, Pant AK, Prakash O, Kumar R, Stocki M, Isidorov VA. Chemical composition, antimicrobial, and antioxidant activities of the essential oils from stem, leaves, and seeds of *Caryopteris foetida* (D. don) Thell.

Indian Journal of Natural Products and Resources. 2021;**12**(2):214-224

[58] Joshi M, Himani Kumar R, Prakash O, Pant AK, Rawat DS. Chemical composition and biological activities of *Nepeta hindostana* (Roth) Haines, *Nepeta graciliflora* Benth. and *Nepeta cataria* L. from India. Journal of Medicinal Herb. 2021;**12**(2):35-46

[59] Thapa P, Prakash O, Rawat A, Kumar R, Srivastava RM, Rawat DS, et al. Essential oil composition, antioxidant, anti-inflammatory, insect antifeedant and sprout suppressant activity in essential oil from aerial parts of *Cotinus coggygria* scop. Journal of Essential Oil Bearing Plants. 2020;**23**(1):65-76

[60] Shahat AA, Ibrahim AY, Hendawy SF, Omer EA, Hammouda FM, Abdel-Rahman FH, et al. Chemical composition, antimicrobial and antioxidant activities of essential oils from organically cultivated fennel cultivars. Molecules. 2011;**16**(2):1366-1377

[61] Han F, Ma GQ, Yang M, Yan L, Xiong W, Shu JC, et al. Chemical composition and antioxidant activities of essential oils from different parts of the oregano. Journal of Zhejiang University-Science B. 2017;**18**(1):79-84

[62] Farouk A, Fikry R, Mohsen M. Chemical composition and antioxidant activity of *Ocimum basilicum* L. essential oil cultivated in Madinah Monawara, Saudi Arabia and its comparison to the Egyptian chemotype. Journal of Essential Oil Bearing Plants. 2016;**19**(5):1119-1128

[63] Tomaino A, Cimino F, Zimbalatti V, Venuti V, Sulfaro V, De PA, et al. Influence of heating on antioxidant activity and the chemical composition of some spice essential oils. Food Chemistry. 2015;**89**:549-554

[64] Butterworth PJ, Warren FJ, Ellis PR. Human α-amylase and starch digestion: An interesting marriage. Starch-Stärke. 2011;**63**(7):395-405

[65] Sani DH, Munna AN, Alam MJ, Salim M, Alam M. Evaluation of α-amylase inhibition and cytotoxic activities of the *Arachis hypogaea* and *Cinnamomum tamala*. Current Nutrition and Food Science. 2021;**17**(3):328-336

[66] Tahir HU, Sarfraz RA, Ashraf A, Adil S. Chemical composition and antidiabetic activity of essential oils obtained from two spices (*Syzygium aromaticum* and *Cuminum cyminum*). International Journal of Food Properties. 2016;**19**(10):2156-2164

[67] Siahbalaei R, Kavoosi G, Shakeri R. In vitro antioxidant and antidiabetic activity of essential oils encapsulated in gelatin-pectin particles against sugar, lipid and protein oxidation and amylase and glucosidase activity. Food Science & Nutrition. 2020;**8**(12):6457-6466

[68] Li L, Wang Q, Yang Y, Wu G, Xin X, Aisa HA. Chemical components and antidiabetic activity of essential oils obtained by hydrodistillation and three solvent extraction methods from *Carthamus tinctorius* L. Acta Chromatographica. 2012;**24**(4):653-665

[69] Fisher K, Phillips C. Potential antimicrobial uses of essential oils in food: Is citrus the answer? Trends in Food Science and Technology. 2008;**19**(3):156-164

[70] Holley R, Patel D. Improvement in shelf-life and safety of perishable foods by plant essential oils and smoke antimicrobials. Food Microbiology. 2005;**22**(4):273-292

[71] Dorman HD, Deans SG. Antimicrobial agents from plants:

Antibacterial activity of plant volatile oils. Journal of Applied Microbiology. 2000;**88**(2):308-316

[72] Chang D, Chen P, Chang S. Antibacterial activity of leaf essential oils and their constituents from *Cinnamomum osmophloeum*. Journal of Ethnopharmacology. 2001;**77**(1):123-127

[73] Karagözlü N, Ergönül B, Özcan D. Determination of antimicrobial effect of mint and basil essential oils on survival of *E. coli* O157:H7 and *S. typhimurium* in fresh-cut lettuce and purslane. Food Control. 2011;**22**(12):1851-1855

[74] Al-Bayati FA. Synergistic antibacterial activity between *Thymus vulgaris* and *Pimpinella anisum* essential oils and methanol extracts. Journal of Ethnopharmacology. 2008;**116**(3):403-406

[75] Raut JS, Karuppayil SM. A status review on the medicinal properties of essential oils. Industrial Crops and Products. 2014;**62**:250-264

[76] Kordali S, Kotan R, Mavi A, Cakir A, Ala A, Yildirim A. Determination of the chemical composition and antioxidant activity of the essential oil of *Artemisia dracunculus* and of the antifungal and antibacterial activities of Turkish *Artemisia absinthium, A. dracunculus, Artemisia santonicum,* and *Artemisia spicigera* essential oils. Journal of Agricultural and Food Chemistry. 2005;**53**(24):9452-9458

[77] Saikia D, Khanuja SP, Kahol AP, Gupta SC, Kumar S. Comparative antifungal activity of essential oils and constituents from three distinct genotypes of *Cymbopogon* spp. Current Science. 2001;**80**(10):1264-1266

[78] Benkeblia N. Antimicrobial activity of essential oil extracts of various onions (*Allium cepa*) and garlic (*Allium sativum*). LWT-Food Science and Technology. 2004;**37**(2):263-268

[79] Arena ME, Alberto MR, Cartagena E. Potential use of citrus essential oils against acute respiratory syndrome caused by coronavirus. Journal of Essential Oil Research. 2021;**33**:1-12

[80] Astani A, Reichling J, Schnitzler P. Screening for antiviral activities of isolated compounds from essential oils. Evidence-Based Complementary and Alternative Medicine. 2011;**2011**:1-8

[81] Romeilah RM, Fayed SA, Mahmoud GI. Chemical compositions, antiviral and antioxidant activities of seven essential oils. Journal of Applied Sciences Research. 2010;**6**(1):50-62

[82] Aktar MW, Sengupta D, Chowdhury A. Impact of pesticides use in agriculture: Their benefits and hazards. Interdisciplinary Toxicology. 2021;**2**:1-12

[83] Isman MB. Plant essential oils for pest and disease management. Crop Protection. 2000;**19**(8-10):603-608

[84] Amaral AC, Ramos AS, Pena MR, Ferreira JL, Menezes JM, Vasconcelos GJ. Acaricidal activity of Derris floribunda essential oil and its main constituent. Asian Pacific Journal of Tropical Biomedicine. 2017;**7**:791-796

[85] Rawat A, Thapa P, Prakash O, Kumar R, Pant AK, Srivastava RM, et al. Chemical composition, herbicidal, antifeedant and cytotoxic activity of *Hedychium spicatum* Sm.: A Zingiberaceae herb. TPR. 2019;**3**(2):123-136

[86] Jiang H, Wang J, Song L, Cao X, Yao X, Tang F, et al. Analysis of essential oils composition from leaves, twigs and seeds of *Cinnamomum camphora* L.

Presl and their insecticidal and repellent activities. Molecules. 2016;**21**:423

[87] Kumar R, Kumar R, Prakash O. Evaluation of in-vitro herbicidal efficacy of essential oil and chloroform extract of *Limnophila indica*. The Pharma Innovation Journal. 2021;**10**:386-390

[88] Jeon YJ, Lee SG, Lee HS. Acaricidal and insecticidal activities of essential oils of *Cinnamomum zeylanicum* barks cultivated from France and India against Dermatophagoides spp., *Tyrophagus putrescentiae* and *Ricania sp*. Applied Biological Chemistry. 2017;**60**:259-264

[89] Sethi S, Prakash O, Chandra M, Punetha H, Pant AK. Antifungal activity of essential oils of some Ocimum species collected from different locations of Uttarakhand. Indian Journal of Natural Products and Resources. 2013;**4**:392-397

[90] Sethi S, Prakash O, Pant AK. Essential oil composition, antioxidant assay and antifungal activity of essential oil and various extracts of *Alpinia allughas* (Retz.) Roscoe leaves. Cogent Chemistry. 2015;**1**(1):1079349

[91] Sethi S, Prakash O, Pant AK. Phytochemical analysis, antioxidant assay and antifungal activity of essential oil and various extracts of *Alpinia malaccensis* (Burm. f.) Roscoe leaves. Cogent Chemistry. 2016;**2**(1):1223781

Chapter 2

# Physiochemical Properties of Essential Oils and Applications

*Sunil Kumar Yadav*

## Abstract

Essential oils have received increasing interest due to the high potential of their novel properties, i.e. antibacterial, antifungal and antioxidant activities. Essential oils are obtained from various parts of aromatic cultures, i.e. roots, leaves, seeds, bark, fruits, flowers, stems, etc. by various oil production methods, i.e. field distillation unit (FDU), steam distillation, water and steam distillation & several advanced (supercritical fluid extraction). Therefore, it is necessary to understand the characterization of the essential oils. This study reports on the method of determination of physiochemical properties with the test parameters, i.e. odor, color, optical rotation, solubility, refractive index, specific gravity, acid value, ester value, and ester value after acetylation. There is also discussion about instruments such as gas chromatography-mass spectrometry due to one of the best tools for identifying and quantifying the constituents of essential oils as its simplicity, rapidity, accuracy, and efficiency.

**Keywords:** essential oils, physiochemical properties, method, components, application

## 1. Introduction

Essential oils are volatile components of aromatic or aromatic crops that give aroma due to their volatility. Generally, aromatic cultures are those that have aromatic compounds that are volatile at room temperature and give a smell [1]. These compounds are present in essential oils preserved in plant cells, tissues, stomas, and other parts of the plant. Usually, essential oils are stored in the roots, leaves, seeds, bark, fruits, flowers, and stems of plants [2]. The essential oil of the various parts of a plant can be obtained by various distillation methods such as hydro-electric distillation, hydro-vapor distillation, steam distillation, solvent extraction, and supercritical extraction, etc. [3]. Essential oils are the secondary plant metabolites synthesized in different parts of the plant, such as leaves, flowers, stems, roots, and seeds [4]. These are of great importance for perfumery and pharmacy. Natural essential oils are considered biodegradable and have no residual toxicity [5]. Due to the improvement in living standards and taste for natural essential oils as fragrant, flavoring, and pharmaceutical ingredients, the demand for natural essential oils has increased in many ways in the recent past. Many industries use synthetic fragrances that are developed in a laboratory to mimic the aromatic and chemical components of natural oils, based on plants that are more expensive to produce. However, synthetic fragrances may not contain the beneficial aspects of natural plant-based essential oils and may even be dangerous for human applications. Chemicals found in artificial fragrances include, for example, phthalates, endocrine disruptors, and carcinogens known as benzene derivatives [6].

On the other hand, the global market for natural fragrances has grown strongly due to the increasing use of natural fragrances such as essential oils over synthetic fragrances as a result of their associated numerous health benefits associated with them, such as aromatherapy, which will drive market growth in the coming years [7]. The analysis of the essential oil purity test can be confirmed with physiochemical, instrumental analysis. The qualitative and quantitative analysis is carried out to know the components of the oil and the percentage of the components contained in the oil respectively, in doing so, we can know the purity of this particular oil. Only pure oils contain a full range of compounds that simply cannot duplicate cheap imitations.

## 2. Physico-chemical analysis

### 2.1 Odor evaluation

Take smelling strips and one end of each odors strip must be clearly marked before use. Now, dip the unmarked end of a strip (about 0.5 to 1.0 ml) in the material under examination and of another strip to the same depth in the standard sample after it has attained room temperature. For certain perfumery materials, such as fatty, absolute, and solid aldehydes, solutions of I to 10 percent solutions in ethyl alcohol or diethyl phthalate for olfactory evaluation [8]. Odorant profile of essential oil can be investigated by gas chromatography (GC)-olfactometry using aroma extract dilution analysis (AEDA) and vocabulary-intensity-duration of elementary odors by sniffing (VIDEO-Sniff) [9]. **Table 1** shows the standards for the odor of some essential oils, according to the Bureau of Indian Standards (BIS).

### 2.2 Solubility determination

Take exactly 1.0 ml of the essential oil in the measuring cylinder (**Figure 1**) and place it in a constant temperature device [10]. Then, maintain the specified temperature as mentioned in the specification. Now, add the dilute alcohol (as specified) for the particular materials. i.e. 60%, 70%, 80% 90% while shaking. Record the volume of the alcohol required for producing a clear solution. Standards of solubility results for some essential oils are shown in **Table 1**.

### 2.3 Optical rotation determination

#### *2.3.1 Polarimeter*

Switch on the light source and wait until full luminosity is obtained (**Figure 2a**) [11]. Then put the blank cell (**Figure 2b**) for normal sample use (100 mm cell) in the cell compartment to get the cell error whether dextro or levo. Now fill up the cell with sample. After that rotate the analyzer knob for alignment with polarized light.

| Name of oil | Physical parameters at 27°C | | | | | | Chemical parameters | | |
|---|---|---|---|---|---|---|---|---|---|
| | Color | Odor | Solubility | Optical rotation | Refractive index | Relative density | Acid value | Ester value | Ester value after acetylation |
| 1. Oil of *Mentha arvensis* | Colorless to light yellow | Strong minty, herbal, cooling sensation | 2.5–3 volumes of ethyl alcohol (70% by volume) | −35° to −45° | 1.456–1.4642 | 0.8773–0.9123 | | 3–15 | |
| 2. Oil of Geranium | Yellowish brown | Strong, rose like with a minty top note | 3 volumes of ethyl alcohol (70% by volume) | −7° to −11° | 1.4630–1.4728 | 0.8824–0.8966 | 10 maximum | 50–76 | 205–230 |
| 3. Oil of Vetiver | Brown to reddish brown viscous liquid | Characteristic and persistent woody aroma | 1–2 volumes in 80% ethanol | +15° to +35° | 1.5150–1.5250 | 0.9850–1.0200 | 35 maximum | 5–16% | 110–165% |
| 4. Oil of Citronella (Java) | Pale yellow to light tan clear | Characteristic citrus grassy with rose undernote | 1–2 volumes of ethanol (80% by volume) | −0.5° to −5° | 1.4624–1.4714 | 0.8743–0.8893 | — | — | — |
| 5. Oil of Eucalyptus globulus | Colorless to pale yellow | Aromatic camphoraceous sharp odor | Equal volume of ethyl alcohol, 80% by volume | −5° to +10° | 1.4580–1.4700 | 1.4561–1.4669 | 0.9050–0.9250 | — | — |
| 6. Oil of Clove bud | Colorless or pale yellow | Sweet fruity, spicy | 2 or more volumes of 70% ethanol | 0° to −2° | 1.5230–1.5310 | 1.0350–1.0570 | — | — | — |
| 7. Oil of Cumin seed | Colorless to pale yellow | Strongly aromatic, spicy | 8 volumes of ethanol (80% by volume) | +1° to +8° | 1.4792–1.5092 | 0.8688–0.9187 | — | — | — |
| 8. Oil of Pine | — | Pleasant, sweet odor reminiscent of terpineols with mild aromatic | — | — | 1.4750–1.4820 | 0.9080–0.9280 | 2% maximum | — | — |
| 9. Oil of Cardamom | Colorless to yellow | Spicy and camphorous | 2–5 volumes of ethanol (70% by volume) | +16° to +41° | 1.4575–1.4605 | 0.9190–0.9360 | 7% maximum | 92–150% | — |
| 10. Oil of Patchouli | Light yellow to reddish brown | Leafy, slightly camphoraccous | 1–10 volumes of ethanol (90% by volume) | −66° to −40° | 1.5020–1.5120 | 0.9480–0.9710 | 4% maximum | 10% maximum | — |

| Name of oil | Physical parameters at 27°C | | | | | | Chemical parameters | | |
|---|---|---|---|---|---|---|---|---|---|
| | Color | Odor | Solubility | Optical rotation | Refractive index | Relative density | Acid value | Ester value | Ester value after acetylation |
| 11. Oil of Sandalwood | Nearly colorless to golden yellow | Pleasant, sweet, woody, and persistent | 5 volumes of ethanol (70% by volume) | −20° to −15° | 1.5000–1.5070 | 0.9635–0.9775 | — | 7% maximum | — |
| 12. Oil of Ginger | Light yellow to greenish yellow liquid | Sharp lemon top note with persistent dry spicy note | In 1 volume of ethanol | −28° to −48° | 1.4860–1.4960 | 0.8700–0.8820 | — | — | — |
| 13. Oil of Palmarosa (var. *Motia*) | Light yellow to yellow mobile | Rosaceous, with grassy background | 2 volumes of ethanol (70% by volume) | −1.4° to +3° | 1.4710–1.4780 | 0.8800–0.8940 | 1% maximum | 7–36% | 260–280% |
| 14. Oil of Lemongrass | Dark yellow to light brown-red | Lemon like | 3 volumes of ethanol (70% by volume) | −3° to +1° | 1.4799–1.4859 | 0.886–0.896 | — | — | — |
| 15. Oil of Basil | — | Spicy herbal, anise like | More than 7 volumes of ethanol (80% by volume) | −6° to +7.5° | 1.4580–1.5480 | 0.9050–0.9620 | 1% maximum | — | — |
| 16. Oil of Cinnamon leaf | — | — | In 2 volumes of ethanol (70% by volume) | 1.5250–1.5360 | 1.0300–1.0560 | — | — | — | — |
| 17. Oil of Dill Seed | — | — | 0.5 or more volumes of ethanol (90% by volume) | +50° to +65° | 1.4750–1.4870 | 0.9360–0.9800 | — | 35–42% | 50–65% |
| 18. Oil of Davana | Clear brownish yellow to brown liquid | Sweet, lingering, fruity, balsamic | 9 volumes of ethanol (90% by volume) | +34° to +41° | 1.4775–1.4995 | 0.9160–0.9560 | 3% max | 30–45% | — |
| 19. Oil of Celery Seed | Pale yellow to light brown | Very persistent and spicy, pleasing | 10 volumes of ethanol (90% by volume) | +50° to +80° | 1.4765–1.4865 | 0.8710–0.9100 | 3.5% (max.) | — | — |
| 20. Oil of Turpentine | — | — | 5 volumes of ethanol (90% by volume), 1 volume of ethanol (95% by volume) | 1.4670–1.4770 | 0.8600–0.8700 | — | 1% (max.) | — | — |

| Name of oil | Physical parameters at 27°C | | | | | | Chemical parameters | | |
|---|---|---|---|---|---|---|---|---|---|
| | **Color** | **Odor** | **Solubility** | **Optical rotation** | **Refractive index** | **Relative density** | **Acid value** | **Ester value** | **Ester value after acetylation** |
| 21. Oil of Himalayan Cedarwood | Pale yellow to deep reddish yellow | Woody, balasamic | In all proportion of ethanol (90% by volume) | +55° to +65° | 1.5050–1.5132 | 0.9280–0.9360 | 1% (max.) | 15% (max.) | 30% (max.) |
| 22. Oil of Black Pepper | Almost colorless to bluish green liquid | Characteristic, recalling that of whole pepper | Completely soluble in 3 volume of ethanol (95% by volume) | −28° to −3° | 1.4730–1.4891 | 0.8450–0.9265 | — | 11% (max.) | — |
| 23. Oil of Jamarosa | Colorless to pale yellow | Sweet floral rosy, citrusy, minty top note | One volume of ethanol (80% by volume) | −2° to +2° | 1.4680–1.4745 | 0.8830–0.8890 | — | 19–33% | — |
| 24. Oil of Rose | Colorless to light yellow | Rosy | IS 15740:2007 (RA 2012) | −2° to −4.5° | 1.4520–1.4680 | 0.8700–0.8800 | — | — | — |

**Table 1.**
*Physico-chemical properties of essential oils [1–17].*

**Figure 1.**
*Measuring cylinder.*

### *2.3.2 Digital polarimeter*

Switch on the power input in the instrument (**Figure 3**) then switch on the light source (sodium vapor lamp) knob of the instrument and wait until the energy (70–80 i.e. full intensity of the lamp) is obtained. Now, check the cell error (dextro or levo) then fill the sample in the cell. Put the cell in the cell compartment of the instrument and note the reading, directly from the display of instrument. Optical oration results of some essential oils are shown in **Table 1**.

## 2.4 Refractive index determination

Sample should be free from moisture and any other residual matters and record the ambient temperature [12]. Then open the prism of ABBE type refractometer (**Figure 4**) and clean it with soft cotton. Now, place some drops of the oil to be tested on the lower part of the prism and close the refractometer. Then, observe through the eyepiece and turn the dispersion correction compensator knob until the colored indistinct boundary seen between the light and darkfield becomes a sharp line. Now, adjust the knurled knob until the sharp line exactly intersects the midpoint of the cross wires in the image. Read the refractive index from the magnifier in the pointer and record the reading. RI results of some essential oils are shown in **Table 1**.

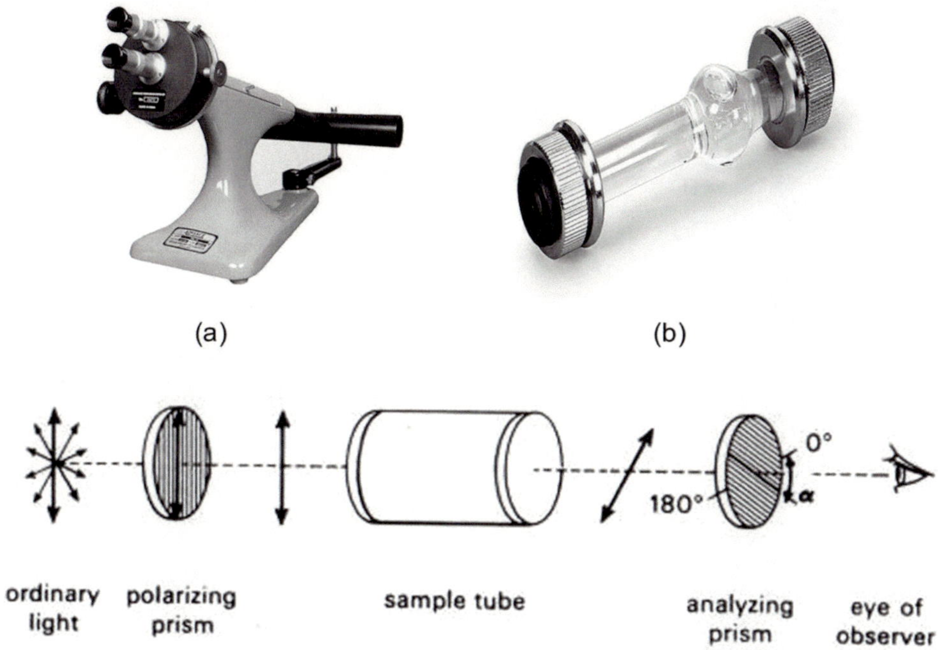


**Figure 2.**
*(a) Polarimeter and (b) sample cell.*

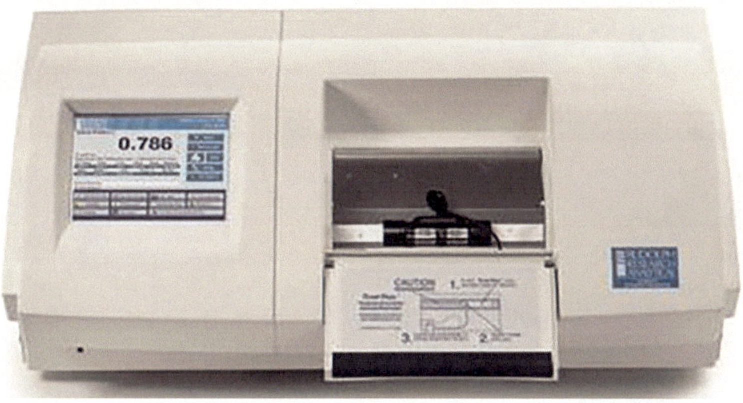


**Figure 3.**
*Digital polarimeter.*

### 2.5 Specific gravity determination

The sample should be free from moisture and any other residual matters [13]. Then, carefully wash and clean the pyknometer (**Figure 5**) or specific gravity bottle and dry the interior with a current of dry air. Now, weigh the pyknometer or specific gravity bottle and record the weight and fill the pyknometer or specific gravity bottle with distilled water and record the weight with temperature. Then again clean the pyknometer/specific gravity bottle and dry the interior with a current of dry air.

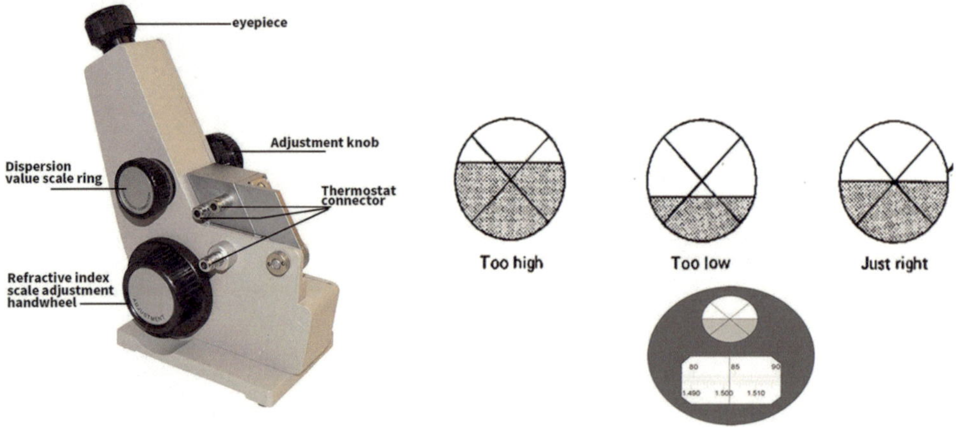


**Figure 4.**
*Abbe type refractometer.*

**Figure 5.**
*Pyknometer/specific gravity bottle.*

Now, again fill the same pyknometer/specific gravity bottle with the material under test and record the weight with temperature.

It can be calculated by the equation as under.

$$d = \frac{W1 - W}{W2 - W} \quad (1)$$

where, d = density, $W$ = weight of pycnometer or relative density bottle, $W1$ = weight of sample, $W2$ = weight of distilled water. The relative densities value of some essential oils is shown in **Table 1**.

### 2.6 Acid value determination

Sample should be free from moisture and any other residual matters [14]. Weigh about 2.5 g of sample material. Then, dissolve the sample in 20 ml rectified spirit (neutralized). Now, titrate it with potassium hydroxide 0.1 N solution (aqueous/alcoholic) using phenolphthalein indicator until the solution remains faintly pink after 10 s of shaking. Now, note the volume of KOH consumed and put the value in the formula.

$$\text{Acid value} = \frac{56.1 \times V \times N}{M} \tag{2}$$

where V = volume of KOH consumed, N = normality of KOH solution, M = weight of material (grams) taken.

### 2.7 Ester value determination

Take an appropriate sample of the material reserved from the acid value determination [15]. Now, add 25 ml of 0.5 N alcoholic KOH, and reflux it on water bath for 1 h. Then, cool it and add 20 ml distilled water and remove the condenser. Now, add few drops of phenolphthalein as an indicator and titrate it against 0.5 N HCl. Simultaneously, a blank determination is also carried out (conditions remain the same except the material to be tested). Put the values in the following formula.

$$\text{Ester value} = \frac{56.1 \times N\,(V1 - V2)}{M} \tag{3}$$

where N = normality of HCl, V1 = vol. in ml of HCl used for blank determination, V2 = vol. in ml. of HCl used in determination to neutralize the excess alkali after hydrolysis, M = weight in g of the material taken.

### 2.8 Ester value after acetylation determination

Take 10 ml of sample and 20 ml acetic anhydride and 2 g of anhydrous sodium acetate in a round bottom flask then add fragments of pumice-stone or porcelain pieces [16]. Now, connect the flask with an air condenser for reflux for 2 h. After this, cool the content. Now, add 50 ml of cold water and heat it at a temperature between 40 and 50°C for 15 min. Then again, cool it and transfer it to a separating funnel. After that, wash the flask twice with 10 ml of distilled water and separate the water layer from the oil layer.

Wash the oil layer by shaking successively with (a) 50 ml of sodium chloride solution (brine solution), (b) 50 ml of sodium carbonate (solution 2% in brine), (c) 50 ml of sodium chloride solution (brine), and (d) 20 ml of distilled water. Now, shake the acetylated sample material vigorously with the distilled water and check the water layer with litmus paper as should be neutral. After that, dry the acetylated sample material by adding anhydrous sodium sulfate for the saponification of the ester hydrolysis process. The saponification process is carried out as under steps needed.

First take 1–1.5 g of acetylated sample material into a flask and add 25 ml of 0.5 N alcoholic KOH solution. Then, reflux it in a water bath for one-hour and cool it then add 20 ml of distilled water from the top of the condenser. Now, titrate it against 0.5 N HCI in the same way, a blank titration is also carried out at the same condition without a sample. Put the value in the following formula for calculation of ester value after acetylation as

$$\text{Ester value after acetylation} = \frac{56.1 \times \text{N} \times (\text{V1} - \text{V2})}{\text{M}} \tag{4}$$

where, N = normality of HCl, $V_1$ = volume in ml of HCl used for blank determination, $V_2$ = volume in ml of HCl used in determination to neutralize the excess alkali after hydrolysis, and M = mass in g of the material taken.

## 2.9 Instrumental analysis

### *2.9.1 GC-MS (gas chromatography-mass spectrometry)*

Gas chromatography-mass spectrometry (GC-MS) analysis is more feasible for the authenticity of essential oil as determining its maximum chemical component. The

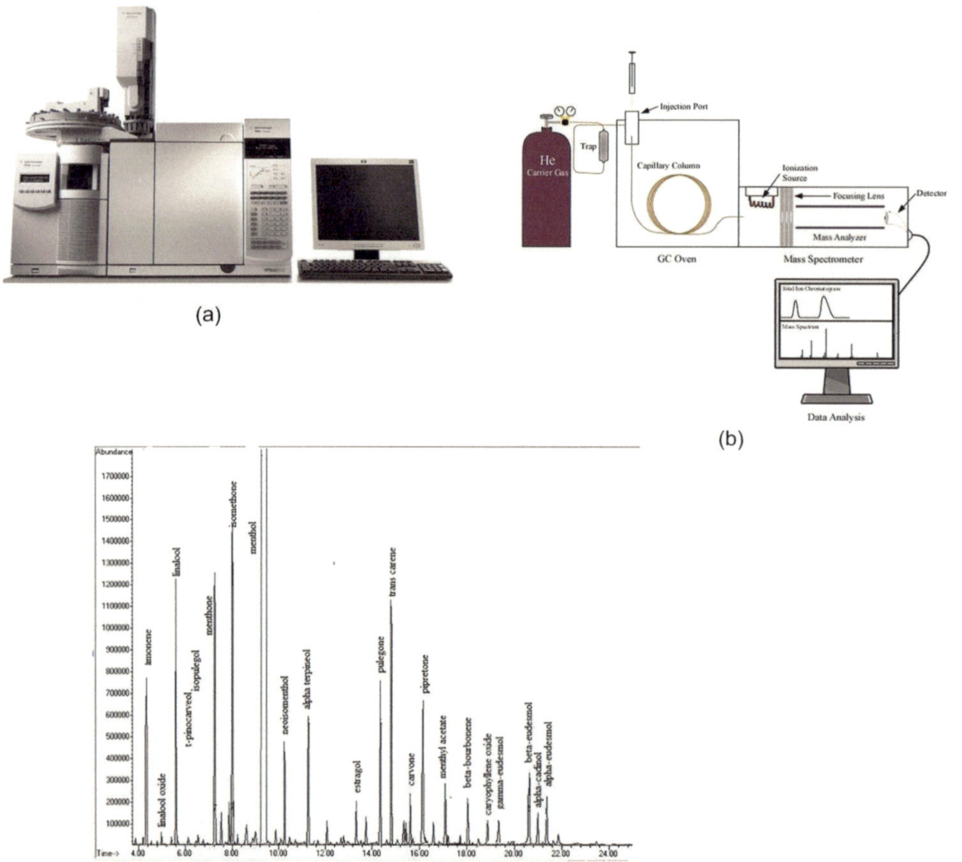


**Figure 6.**
*(a) Gas chromatography-mass spectrometry instrument, (b) role of the different parts of gas chromatography-mass spectrometry, and (c) gas chromatography-mass spectrometry-chromatogram.*

qualitative GC-MS analysis for the extracted essential oils can be carried out by using HP 6890 gas chromatography coupled with HP 5973 mass selective detector (**Figure 6a**) operating in 70 eV mode. Samples of 0.2 μL need to inject in the capillary column with the split mode at a ratio of 5:1. The compounds separate on a 30 m long capillary column (HP-5MS), 0.25 mm in diameter, and with 0.25 μm thick stationary phase film (5% phenyl)-methylpolysiloxane). The flow rate of helium into the column needs to be kept at 1.2 mL $min^{-1}$. Initially, the temperature of the column 45°C, then it increases to 200°C at a rate of 5°C $min^{-1}$ (kept constant for 10 min), and then heat up to a final temperature of 250°C at a rate of 5°C $min^{-1}$ (**Figure 6b**).

The oven stays kept at this temperature for 20 min. The solvent delays 4 min. The total running time for a sample is about 70 min. The relative percentage of the essential oil constituents evaluate from the total peak area (TIC) by apparatus software [18]. Essential oil constituents' identification by comparison of their mass spectra (**Figure 6c**) with those stored in the libraries i.e. NIST (National Institute of Standards and Technology), flavor, and Adam's mass spectral libraries using various search engines. **Table 2** is an example of some essential oils and their major components.

| Name of essential oil | Major chemical constituents of molecule | Reference |
|---|---|---|
| 1. Oil of *Mentha arvensis* | Menthol (84.63%), L-menthol (4.58%) | [19] |
| 2. Oil of Geranium | Citronellol (37.5%), geraniol (6.0%), caryophyllene oxide (3.7%), menthone (3.1%), linalool (3.0%), β-bourbonene (2.7%), iso-menthone (2.1%) and geranyl formate (2.0%) | [20] |
| 3. Oil of Vetiver | Vetiverol (45–80%), khusimol (3.4–13.7%), β-vetispirene (1.6–4.5%), vetiselinenol (1.3–7.8%), vetivone (2.5–6.3%) | [21] |
| 4. Oil of Citronella (Java) | Citronellal (29.6%), 2,6-octadienal, 3,7-dimethyl-, (E)-(11%), cis-2,6-dimethyl-2,6-octadiene (6.9%), caryophyllene (6.5%), citronellol (4.8%), limonene (2.7%) | [22] |
| 5. Oil of *Eucalyptus globulus* | Eucalyptol (51.62%), α-pinene (23.62%), p-cymene (10%), β-myrcene (8.74%), terpinen-4-ol (2.74%) and γ-terpinene (2.59%) | [23] |
| 6. Oil of Clove bud | Eugenol (76.8%), followed by β-caryophyllene (17.4%), α-humulene (2.1%), and eugenyl acetate (1.2%) | [24] |
| 7. Oil of Cumin seed | Cuminaldehyde (36.67%) and caren-10-al (21.34%) | [25] |
| 8. Oil of Pine | α-Terpineol (30.2%), linalool (24.47%), limonene (17.01%), anethole (14.57%), caryophyllene (3.14%), and eugenol (2.14%) | [26] |
| 9. Oil of Cardamom | α-Terpinyl acetate (29.9–61.3%) followed by 1,8-cineole (15.2–49.4%), α-terpineol (0.83–13.2%), β-linalool (0.44–11.0%), and sabinene (1.9–4.9%) | [27] |
| 10. Oil of Patchouli | Patchouli alcohol (42.75%), Delta-Guaiene (28.30), Azulene (20.48%), Trans Caryophyllene (11.84%), Seychellene (CAS) (10.77%), Nephtalene (8.02%), Cycloheptane (6.02%) and Caryophyllene (5, 73%) | [28] |
| 11. Oil of Sandalwood | α-Santalol (59.00%), α-bergamotene (9.68%), and β-santalol (9.02%) | [29] |
| 12. Oil of Ginger | a-Zingiberene (30.06%), β-sesquiphellandrene (10.71%), E–E-a-farnesene (9.75), β-bisabolene (6.53%), y-curcumene (5.90%) and ar-curcumene (5.18%) | [30] |

| Name of essential oil | Major chemical constituents of molecule | Reference |
|---|---|---|
| 13. Oil of Palmarosa (var. *Motia*) | (E)-β-Ocimene (1.2–4.3%), linalool (0.8–2.0%), geraniol (70.1–85.3%), geranyl acetate (4.3–14.8%) and (E,Z)-farnesol (1.6–3.4%) | [31] |
| 14. Oil of Lemongrass | Citral-a (33.1%), citral-b (30.0%), geranyl acetate (12.0%) and linalool (2.6%) | [32] |
| 15. Oil of Basil | Methyl cinnamate (70.1%), linalool (17.5%), β-elemene (2.6%) and camphor (1.52%) | [33] |
| 16. Oil of Cinnamon leaf | Eugenol (74.9%), followed by β-caryophyllene (4.1%), benzyl benzoate (3.0%), linalool (2.5%), eugenyl acetate (2.1%) and cinnamyl acetate (1.8%) | [34] |
| 17. Oil of Dill Seed | Carvone (38.9%), apiol (30.8%), limonene (15.9%) and trans-(+)-dihydrocarvone (10.9%) | [35] |
| 18. Oil of Davana | cis-Davanone (45.8%), bicyclogermacrene (9.6%), linalool (2.5%), caryophyllene oxide (2.2%) and phytol (2.1%) | [36] |
| 19. Oil of Celery Seed | Limonene (54.04–58.29%), myrcene (19.51–27.65%), 1,2 ethanediol, 1-phenyl (5.62–7.17%) | [37] |
| 20. Oil of Turpentine | α-Pinene 77% and 89% respectively as the major component | [38] |
| 21. Oil of Himalayan Cedarwood | β-Himachalene (38.3%), α-himachalene (17.1%) and γ-himachalene (12.6%) | [39] |
| 22. Oil of Black Pepper | trans-Caryophyllene (30.33%), limonene (12.12%) | [40] |
| 23. Oil of Jamarosa | Geraniol (80–90%), geranyl acetate (19–33%) | [41] |
| 24. Oil of Rose | Citronellol (15.9–35.3%), geraniol (8.3–30.2%), nerol (4.0–9.6%), nonadecane (4.5–16.0%), heneicosane (2.6–7.9%) and linalool (0.7–2.8%) | [42] |

**Table 2.**
*Major constituents of the essential oils.*

## 3. Applications

The applications of essential oils are diverse. Widely used in cosmetics and perfumes, they also have medicinal applications due to their therapeutic properties as well as agro-alimentary uses because of their antimicrobial and antioxidant effects.

a. Oil of *Mentha arvensis*

Uses for stomach disorders, inflammation, and treatment of fever headache, cold, and asthma.

b. Oil of Geranium

Uses for female reproductive disorders, menstrual cramps, infertility, endometriosis, premenstrual syndrome. Menopausal symptoms, circulatory disorders, Raynaud's disease, varicose veins, hemorrhoids, neuralgia, nervous skin disorders, depression, fatigue, emotional crisis, stress-related conditions, wounds, acne, bruises, minor burns, dermatitis, eczema, ulcers, hemorrhoids, head lice, ringworm, sebum balancing, urinary, and liver tonic.

c. Oil of Vetiver

Uses for nervous tension, muscular spasm, muscular pain, menstrual cramps, premenstrual syndrome, restlessness, acne, arthritis, cuts, depression, exhaustion, insomnia, muscular aches, oily skin, rheumatism, sores, stress.

d. Oil of Citronella (Java)

Uses for muscular aches, infectious skin conditions, fevers, heat rash, excessive perspiration, fungal infections, fatigue, insect bites, insect deterrent.

e. Oil of Eucalyptus globules

Uses for respiratory infection, bronchitis, infectious disease, fever, catarrh, sinusitis, fever, muscular aches and pains, rheumatism, arthritis, urinary infection, cystitis, parasitic infection.

f. Oil of Clove bud

Uses for cognitive support and brain health, pain relief, bacterial infection, fungal infection, viral skin infection, warts, verrucas, toothache, gum disease, muscle pain, rheumatism, flu, bronchitis, tired limbs, nausea, flatulence, stomach cramp, abdominal spasm, parasitic, infection, scabies, ringworm.

g. Oil of Cumin seed

Uses for toxin buildup, poor circulation, low blood pressure, colic, stomach cramps, indigestion, gas, fatigue.

h. Oil of Cardamom

Uses for appetite (loss of), colic, fatigue, stress.

i. Oil of Patchouli

Uses for treating skin conditions such as dermatitis, acne, or dry, cracked skin, easing symptoms of conditions like colds, headaches, and stomach upset, relieving depression, providing feelings of relaxation and helping to ease stress or anxiety, helping with oily hair or dandruff, controlling appetite, using as an insecticide, antifungal, or antibacterial agent, using as an additive in low concentrations to flavor foods like candies, baked goods, and beverages.

j. Oil of Sandalwood

Uses for bronchitis, chapped skin, depression, dry skin, laryngitis, leucorrhea, oily skin, scars, sensitive skin stress, stretch marks.

k. Oil of Ginger

Uses for aching muscles, arthritis, nausea, indigestion, poor circulation, nervous exhaustion.

l. Oil of Palmarosa (var. Motia)

Uses for sinusitis, excess mucus, cystitis, urinary tract infection, gastrointestinal disorders, scarring, wounds acne, pimples, boils, fungal infection, general fatigue, muscular aches, over-exercised muscles, stress, irritability, restlessness, insect bites, and stings.

m. Oil of Lemongrass

Uses for muscular aches and pains, gastrointestinal disorders, indigestion, physical and mental exhaustion acne, insect repellent.

n. Oil of Basil

Uses for bronchitis, colds, coughs, exhaustion, flatulence, flu, gout, insect bites, insect repellent, muscle aches rheumatism, sinusitis.

o. Oil of Cinnamon leaf

Uses for sluggish digestion, colds/flu exhaustion, lice, circulation, rheumatism, scabies, stress.

p. Oil of Dill seed

Uses for dyspepsia, flatulence, indigestion, bronchial asthma, dysmenorrhea, and the promotion of lactation.

q. Oil of Davana

Uses for bacterial infection, bronchial congestion, coughs, colds, influenza, nervous stomach, indigestion, nausea, menstrual cramps, menopausal symptoms, general debility, anxiety, stress, irritability, tension, anxiety, wound healing, antiseptic, coughs.

r. Oil of Himalayan Cedarwood

Uses for acne, arthritis, bronchitis, coughing, cystitis, dandruff, dermatitis, stress.

s. Oil of Black Pepper

Uses for aching muscles, arthritis, chilblains, constipation, muscle cramps, poor circulation, sluggish digestion, quitting smoking, and nicotine addiction.

t. Oil of Jamarosa

Jamarosa essential oil regulates skin moisture & sebum production. Can restore luster to dull aged skin and remove wrinkles & other signs of aging. Has disinfectant, antiseptic properties, and is widely used for treating insect bites and as insect repellents. Aids in repairing the damaged skin cells. Fights anxiety, stress and promotes peaceful sleep.

u. Oil of Rose

Uses for depression, eczema, frigidity, mature skin, menopause, and stress.

## 4. Conclusion

This study will facilitate the identification of essential oils or fragrant raw materials purity and quality. Physiochemical methods and their range of value can be utilized by traders of fragrant raw material as avoiding adulteration. GC-MS is an ideal instrumental analysis for maximum major and minor chemical constituents of the essential oils. Advanced analytical techniques for the characterization of essential oils are more reliable by their fruitful results. Since, good quality of the raw material i.e. essential oils can be used in various purposes i.e. antimicrobial, insecticide, antiseptic, antifungal, and analgesic activities, aromatherapy, disease treatments and cosmetics & allied products as desired results. Quality assessment improves the confidence of both producer & consumer as well.

## Conflict of interest

The authors have no conflict of interest to declare.

## Author details

Sunil Kumar Yadav
Quality Assessment Laboratory, Fragrance and Flavor Development Centre, Ministry of MSME, Government of India, Kannauj, U.P., India

*Address all correspondence to: sunilffdc@gmail.com

## References


[1] Wissal D, Sana B, Sabrine J, Nada B, Wissem M. Essential oils' chemical characterization and investigation of some biological activities: A critical review. Medicines (Basel). 2016; **3**(4):25. DOI: 10.3390/medicines 3040025

[2] Ghaderinia P, Shapouri R. Assessment of immunogenicity of alginate microparticle containing *Brucella melitensis* 16M oligo polysaccharide tetanus toxoid conjugate in mouse. Banat's Journal of Biotechnology. 2017; **8**(16):83-92

[3] Qing-Wen Z, Li-Gen L, Wen-Cai Y. Techniques for extraction and isolation of natural products: A comprehensive review. Chinese Medicine. 2018;**13**:20

[4] Jones M, Kossel A. A biographical sketch. Yale Journal of Biology and Medicine. 1953;**26**(1):80-97

[5] Falleh H, Jemaa MB, Saada M, Ksouri R. Essential oils: A promising eco-friendly food preservative. Food Chemistry. 2020;**330**:127268. DOI: 10.1016/j.foodchem.2020.127268

[6] Olujimi O, Fatoki OS, Odendaal JP, Okonkwo J. Endocrine disrupting chemicals (phenol and phthalates) in the South African environment: A need for more monitoring. Water SA. 2010;**6**:5. DOI: 10.4314/wsa.v36i5.62001

[7] Jugreet BS, Fawzi MM, Gokhan Z, Filippo M. Essential oils as natural sources of fragrance compounds for cosmetics and cosmeceuticals. Molecules. 2021;**26**(3):666. DOI: 10.3390/molecules26030666

[8] Bureau of Indian Standards (BIS) method for olfactory assessment of natural and synthetic perfumery materials (First Revision). IS 2284: 1988 (1 Revision): Reaffirmation 2019

[9] Katharina B, Pascal T, Xavier F, Uwe JM, Hugues B, Daniel J, et al. Identification of odor impact compounds of *Tagetes minuta* l. essential oil: Comparison of two GC-olfactometry methods. Journal of Agricultural and Food Chemistry. 2009;**57**(18):8572-8580. DOI: 10.1021/jf9016509

[10] Bureau of Indian Standards (BIS). Determination of solubility. IS 326 (Part 6): 2005 (RA 2015)

[11] Bureau of Indian Standards (BIS). Methods of sampling and test for natural and synthetic perfumery materials: Part 4 determination of optical rotation (Third Revision), IS 326: Part 4: 2005/ ISO 592: 1998 (2 Revision)

[12] Methods of sampling and test for natural and synthetic perfumery materials: Part 5 determination of refractive index (Third Revision): IS 326: Part 5:2006/ISO 280: 1998:(3 Revision)

[13] Bureau of Indian Standards (BIS). Methods of sampling and test for natural and synthetic perfumery materials: Part 3 determination of relative density (Third Revision), IS 326:Part 3:2006/ISO 279:1998 (3 Revision)

[14] Bureau of Indian Standards (BIS). Methods of sampling and test for natural and synthetic perfumery materials: Part 7 determination of acid value (Third Revision): IS 326:Part 7:2006/ISO 1242: 1999 (3 Revision)

[15] Bureau of Indian Standards (BIS). Methods of sampling and test for natural and synthetic perfumery materials part8 determination of ester value (Third

Revision): IS 326:Part 8:2005/ISO 709: 2001 (2 Revision)

[16] Bureau of Indian Standards (BIS). Methods of sampling and test for natural and synthetic perfumery materials: Part 9: determination of ester value and free alcohols. Section 1: determination of ester values, before and after acetylation, and evaluation of the contents of free and total alcohols (Third Revision): IS 326:Part 9:Sec 1:2017/ISO 1241: 1996

[17] Bureau of Indian Standards (BIS). Manak Bhavan, 9 Bahadur Shah Zafar Marg, New Delhi-110002. Available from: https://standardsbis.bsbedge.com/

[18] Hassanpouraghdam MB, Hassani A, Shalamzari MS. Menthone-and estragole-rich essential oil of cultivated *Ocimum basilicum* L. from Northwest Iran. Chemija. 2010;**21**:59-62

[19] Harpreet K, Ritu T, Anu K, Charu M. Chemical composition and antifungal activity of essential oils from aerial parts of *Mentha piperita* and *Mentha arvensis*. International Journal of Pharmacognosy. 2018;**5**(12):767-773

[20] Farukh SS, Hanjing Z, William NS. Composition of geranium (*Pelargonium graveolens*) essential oil from Tajikistan. American Journal of Essential Oils and Natural Products. 2014;**2**(2):13-16

[21] Smitha GR, Varghese TS, Manivel P. Vol. 387. Anand, Gujarat: Icar—Directorate of Medicinal and Aromatic Plants Research Boriavi; 2014. p. 310

[22] Lee SW, Wendy W. Chemical composition and antimicrobial activity of cymbopogon nardus citronella essential oil against systemic bacteria of aquatic animals. Iranian Journal of Microbiology. 2013;**5**(2):147-152

[23] Ismail A, Ester I, Francis M, William K. Chemical composition of essential oils from Eucalyptus globulus and Eucalyptus maculata grown in Tanzania. Scientific African. 2021;**12**:e00758

[24] Nurdjannah N, Bermawie N. Handbook of Herbs and Spices. Woodhead Publishing Series in Food Science, Technology and Nutrition; 2012. pp. 197-215

[25] Sahadeo DP, Maniker PP, Sushilkumar JW, Chandrakiran SU, Mahendra KR. Chemical composition, antimicrobial and antioxidant activity of essential oils from cumin and ajowan. 2016;**8**(1):60-65

[26] Wei-Cai Z, Zeng Z, Hong G, Li-Rong J, Qiang H. Chemical composition, antioxidant, and antimicrobial activities of essential oil from pine needle (*Cedrus deodara*). Journal of Food Science. 2012; **77**:C824-C829

[27] Kaliyaperumal AK, Sampathrajan V, Muthusamy M, Dhanya MK, Gunasekaran A, Shaji A, et al. Essential oil profile diversity in cardamom accessions from southern India. Frontiers in Sustainable Food Systems. 2021;**5**:1-8

[28] Ermaya D, Sari SP, Patria A, Hidayat F, Razi F. Identification of patchouli oil chemical components as the results on distillation using GC–MS. IOP Conference Series: Earth and Environmental Science. 2019;**365**:1-5

[29] Upul S, Manuri G, Hettiarachchi DS. Essential oil content and composition of Indian sandalwood (*Santalum album*) in Sri Lanka. Journal of Forest Research. 2013;**24**:27-130

[30] Pantea S, Roshanak M, Leila A, Maryam HM. Volatile constituents of ginger oil prepared according to Iranian

traditional medicine and conventional method: A comparative study. African Journal of Traditional, Complementary, and Alternative Medicines. 2016;**13**(6): 68-73

[31] Rajeswara Rao BR, Rajput DK, Patel RP. Essential oil profiles of different parts of palmarosa (*Cymbopogon martinii* (roxb.) wats. var. motia burk.). Journal of Essential Oil Research. 2009;**21**:519-521

[32] Somit RC, Tandon PK, Chowdhury AR. Chemical composition of the essential oil of *Cymbopogon flexuosus* (steud) wats. growing in Kumaon region. Journal of Essential Oil-Bearing Plants. 2010;**13**:588-593

[33] Poonkodi K, Subban R. Chemical composition of the essential oil from basil (*Ocimum basilicum* Linn.) and its in vitro cytotoxicity against HeLa and HEp-2 human cancer cell lines and NIH 3T3 mouse embryonic fibroblasts. Natural Product Research. 2012;**26**(12):1112-1118

[34] Erich S, Leopold J, Gerhard B, Gernot AE, Ivanka S, Albert K, et al. Composition and antioxidant activities of the essential oil of cinnamon (*Cinnamomum zeylanicum* blume) leaves from Sri Lanka. Journal of Essential Oil-Bearing Plants. 2006;**9**:170-182

[35] Vishaldeep K, Ramandeep K, Urvashi B. A review on dill essential oil and its chief compounds as natural biocide. 2021;**36**:412-431

[36] Stefanie B, Gerhard B, Erich S, Juergen W, Alexander S, Albena S, et al. GC-MS-analysis, antimicrobial activities and olfactory evaluation of essential davana (*Artemisia pallens* wall. ex dc) oil from India. Natural Product Communications. 2008;**3**(7):1057-1062

[37] Ewa R, Renata NW, Andrzej S, Piotr G. The chemical composition of the essential oil of leaf celery (*Apium graveolens* l. var. *Secalinum alef*.) under the plants' irrigation and harvesting method. Acta Scientiarum Polonorum Hortorum Cultus. 2016;**15**(1):147-157

[38] Mohamed G, Badr S, Maroc AC, Abderrahman A. Turpentine chemical composition and antimicrobial activity of *Pinus pinaster* and *Pinus halepensis* of Morocco. Acta Botanica Gallica. 2007; **154**:29-300

[39] Chaudhary A, Kaur P, Singh B, Pathania V. Chemical composition of hydrodistilled and solvent volatiles extracted from woodchips of Himalayan Cedrus: *Cedrus deodara* (Roxb.) Loud. Natural Product Communications. 2009; **4**(9):1257-1260

[40] Milos N, Dejan S, Jasmina G, Ana C, Tatjana M, Marija S, et al. Could essential oils of green and black pepper be used as food preservatives? Journal of Food Science and Technology. 2015;**52**: 6565-6573

[41] Nadeem A, Saxena BK. Isolation of geraniol content from various essential oils. The Asian Journal of Experimental Chemistry. 2009;4(1&2):14-17

[42] Ram SV, Rajendra CP, Amit C, Anand S, Ajai KY. Volatile constituents of essential oil and rose water of damask rose (*Rosa damascena* mill.) cultivars from North Indian hills. Natural Product Research. 2011;**25**:1577-1584

Chapter 3

# Essential Oil as Green Preservative Obtained by Ecofriendly Extraction Techniques

*Nashwa Fathy Sayed Morsy*

## Abstract

Essential oils are formed by a complex matrix of substances that are biosynthesized in the secondary metabolism of plants. Nowadays, different ecofriendly extraction techniques (e.g., ultrasound-, microwave-, enzyme-assisted extraction, and supercritical fluid by $CO_2$, etc.) have been adopted to obtain essential oils. These techniques provide unique quality of essentials oils or extracts from aromatic plants in a short time with high energy savings. Essential oils not only impart aroma, but also possess antimicrobial and antioxidant activities. Health limitations in the use of synthetic additives have drawn researchers' attention towards essential oils as safe natural preservatives. Therefore, this chapter summarizes novel technologies to recover essential oils or extracts. In addition, it focuses on application of essential oils and their constituents as green preservatives to retard microbial growth and oxidative spoilage.

**Keywords:** essential oil, green preservative, ecofriendly techniques, enzyme assisted extraction, ultrasound assisted extraction

## 1. Introduction

Conventional techniques (hydrodistillation, steam distillation) used for EO extraction from aromatic plants have several disadvantages such as long extraction time, high energy consumption and degradation of thermally labile aromatic compounds [1]. Characteristic natural flavor of an essential oil depends mainly on its components and their concentrations. Therefore, extraction procedure has to be sensitive enough to keep the proportions of its constituents in their natural state. The oxygenated compounds of EO is considered the main indicator for its quality [2]. The level of these compounds in the EO is affected by the extraction technique used. Recently, novel (green) techniques have been applied solely or in combination with other technique to recover essential oils with high quality in a short time [3]. These green techniques include: Enzyme (EAE), ultrasound (UAE), and microwave (MAE) and supercritical fluid extraction (SFE) [4].

Essential oils (EOs) are used in a wide range of food types as biopreservatives according to their antioxidant and antimicrobial properties [5]. Essential oils are directly added to food matrix, incorporated into food packaging materials, applied in edible coatings or in modified atmosphere packaging [6]. Some EOs and their components such as carvacrol, carvone, cinnamaldehyde, citral, eugenol, linalool, limonene, thymol and vanillin, were accepted for use as flavorings and food additives by the European Commission. Application of essential oils at high concentration required in food preservation is limited by its sensorial characteristics, which would affect negatively the original organoleptic properties of foods [7–9].

## 2. Ecofriendly extraction techniques

### 2.1 Ultrasonic assisted extraction (UAE)

Ultrasonication is considered as a green extraction technique for EO [10]. Ultrasound (U) waves are successfully utilized in the extraction of essential oils, oleoresins, and other bioactive compounds from spice matrices [11]. The advantages of this technique include: low-temperature, short extraction time, low energy consumption, and superior quality of the extracted EO [12]. Microsecond pulses of ultrasound wave generated vapor bubbles within the liquid medium. The bubbles expanded to a large size during expansion cycles before implosion on the surface of plant material, that lead to micro cracking in the cell wall which helps to penetrate the solvent in the cell wall of the powder and release the intracellular components into the medium [13]. This observation could be noticed in **Figure 1**.

There are two types of UAE of EO: ultrasound (US) as a pretreatment prior to HD and simultaneous ultrasound-assisted HD extraction [14]. The UAE efficiency of EO is affected by power, sonication time, frequency, temperature, solvent type and liquid to solid ratio.

Lilia et al. [15] evaluated the effect of UAE time (10, 20, 30, 45 and 60 min) prior to HD on the EO yield from dried flowering tops of *Lavandula stoechas* L. Plants were collected from two regions: Keddara and Adekar in Algeria. The highest yield was obtained in the Adekar sample (1.59%) and Keddara sample (0.87%) after 10 min and 45 min of ultrasound (US), respectively. UAE pretreatment was followed by HD for 90 min. However, the yield of EO obtained by conventional HD (180 min) represented ~70% of that produced by US-HD technique.

Wu et al. [16] extracted EO from dried aerial parts of *Artemisia annua* by ultrasonic-assisted steam distillation extraction. The investigated independent factors were steam distillation (SD) time (1 h, 3.5, and 6 h), US time (0 h, 0.5, 1 h), and solid to liquid ratio (g/mL) (1:6, 1:10 and 1:14). The EO yield reached 0.71% with the optimal conditions (SD extraction time of 3.5 h, US time of 0.5 h, and solid to liquid ratio of 1:10 (g/mL) instead of 0.49% obtained with the conventional SD for 6 h.

Jadhav et al. [17] investigated the UAE of EO from *Piper betle* leaf powder in presence of water at different sonication time (20, 30 and 40 min), dissipated energy (5.64, 12.24, 19.8, 34.56, and 47.32 W) and temperature (30, 40, 50 and 60°C), with different leaf powder to solvent ratios (1:3, 1:4, 1:5 and 1:6). The maximum yield of EO (0.5%) was recorded at 30 min of US irradiation while it did not exceed 0.35% after 3 h of HD. Increasing temperature more than 30°C, dissipated energy higher than 34.56 W and solvent to solid ratio higher than 5 did not significantly increase the yield of EO. They attributed the obtained results at high US power to the formation of large bubble cloud in the solvent that shielded and scattered the bubble energy discharged during the collapse process.

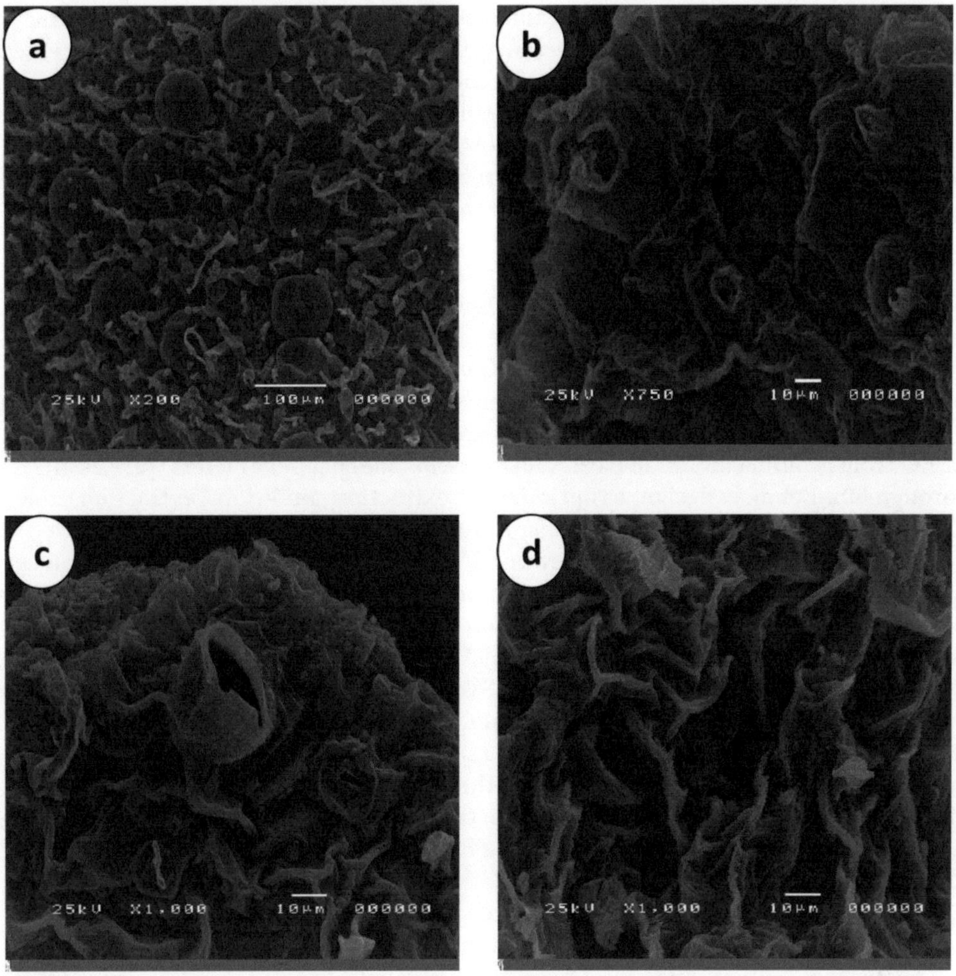


**Figure 1.**
*Scanning electron micrographs of wild mint (Mentha longifolia) leaves (a) control without any treatment (b) after extraction with HD for 3 h (c) after extraction with U (60 W) for 10 min (d) after extraction with U (60 W) for 10 min + HD for 33 min (×200–×1000 magnification, 25 kV).*

Chen et al. [14] studied the effect of the US power on EO yield from cinnamon bark. The samples were subjected to US irradiation ranged from 100 to 500 W for 30 min prior HD. The EO yield (2%) increased with the increase of US power to 300 W. The yield decreased at higher US irradiation (>300 W). They attributed this decrement to high temperature generated, high acoustic pressure created, formation of big bubbles and increase of cavitation bubbles bursting that led to decomposition of EO constituents. Extending US time to 30°min caused a significant increase in EO yield and decrease HD time from 2 h to 1 h. Further increase in US time caused lower yield due to loss of volatile constituents. US at water to solid ratios ranged from 4 to 12 (mL/g) was evaluated. The highest EO yield was recorded for water to solid ratio of 6 (mL/g) and decreased with the continual increase of liquid to solid ratio due to the reduction of the ultrasound intensity needed for the breakage of cell walls. Statistical analysis of the data showed that the order of influence of the dependent factors on EO yield was US time > HD time > US power > liquid to solid ratio.

Ghule et al. [18] extracted eugenol and eugenol acetate from ground clove buds by ultrasound assisted hydrotropic extraction (UAHE). Sodium cumene sulfonate was used to prepare the hydrotropic solution, since solubility of eugenol in water was found to be 1.35 g/L instead of 500 g/L in the aqueous solution (1.8 M) of sodium cumene sulfonate. The extraction time (15–75 min), temperature (30–70°C), hydrotrope concentration (0.2–1.8 M), solid loading (6–22 g/150 mL of hydrotrope solution) and US power (120–200 W) were selected as independent variables. The combined yield of eugenol and eugenol acetate was used as a response. After sonication the mixture was filtered. Eugenol and its derivative were recovered from the filtrate by extraction with hexane. The highest extraction yield (20.04%) was obtained with the following optimal conditions; US power of 158 W, 38°C, hydrotrope concentration of 1.04 M, solid loading of 8.2 g, and extraction time of 30 min. The EO yield obtained by conventional hydrotropic extraction for 1 h was not significantly different than that resulted from UAHE technique.

Guo et al. [19] used ultrasound to enhance subcritical water extraction (USWE) of EO from ground cinnamon bark. The following independent factors were used, through Box-Behnken design, to optimize the extraction conditions: extraction time (20, 25, and 30 min), extraction temperature (120, 130, and 140°C), and US power (100, 125, and 150 W) with a pressure of 5 MPa. The yield of cinnamaldehyde was set as the dependent variable, while its content in the EO was used as a quality index. They compared USWE with the following extraction techniques: steam distillation for 4 h, ultrasound assisted extraction (UAE) by dichloromethane with US power of 150 W for 40 min, and subcritical water extraction (SWE) under pressure of 5 MPa, with a liquid to solid ratio of 12 mL/g, at 132°C for 38 min. Although UAE resulted in the highest EO yield (2.1%) compared to other extraction techniques (1.58–1.83%), the cinnamaldehyde content that obtained with UAE was the lowest (8.965 mg/g). The optimal conditions of USWE were found to be extraction time of 25 min, extraction temperature of 140°C, and US power of 145 W, a pressure of 5 MPa and liquid to solid ratio of 8 mL/g. Under these conditions the maximum yield of EO and cinnamaldehyde content were 1.78% and 12.662 mg/g, respectively. Results indicated that coupling ultrasound with SWE shortened the extraction time and improved the quality of the obtained EO.

Zhang et al. [20] extracted EO from dried citronella leaves with ultrasonic ohmic heating distiller in presence of distilled water. The EO yield of 18 mL/kg dry weight as obtained with liquid to solid ratio of 6. Increasing liquid to solid ratio to 12 caused significant decrease in the yield. They attributed this to difficult in distilling out the EO from large volume of solvent. Meanwhile, increasing US power from 36 to 144 W was accompanied by a significant increase in the EO yield. No significant increase in the EO yield was noticed by further increase in US power. Increasing the current intensity from 1 to 5 A significantly increased EO yield to 20 mL/kg dry weight. This yield was obtained when extraction process was conducted for 40 min, after which no further increase was observed. They suggested that excessive extraction time under the US and ohmic conditions used affected negatively the release of EO due to gelatinization of intracellular components. They used response surface method to optimize the extraction conditions. The EO yield reached 22.91 mL/kg dry weight under the optimal conditions (liquid to solid ratio of 7 mL/g, US power of 180 W, current of 5°A, and time of 30 min).

### 2.2 Enzyme-assisted extraction (EAE)

The EAE technique is considered as a green extraction technology [21]. In EAE, hydrolytic enzymes act on the polysaccharides of the cell wall, disrupt it and release

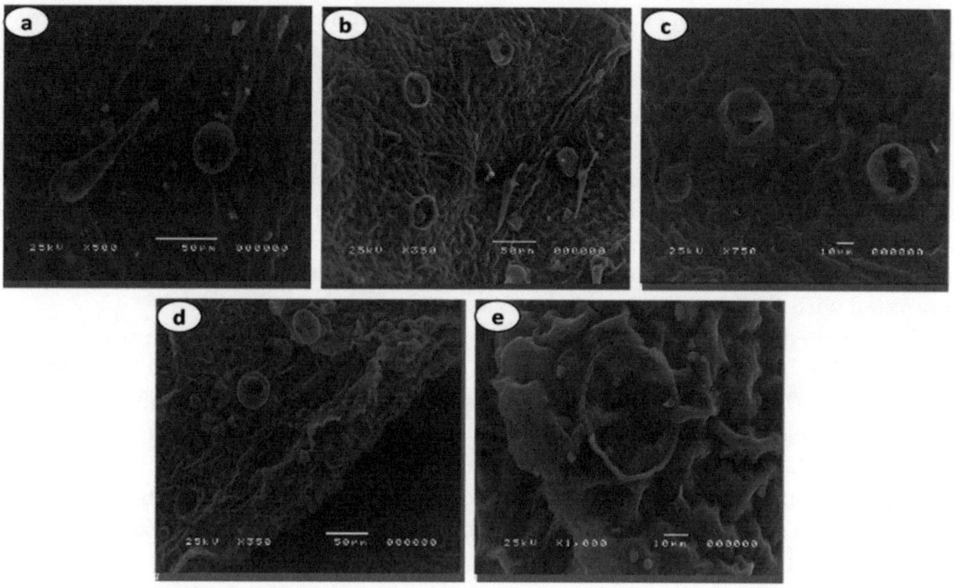


**Figure 2.**
*Scanning electron micrographs of lemon verbena (Aloysia citrodora) leaves (a) control without any treatment (b) after HD (c) pretreated with Cellulase (d) pretreated with pectinase (e) pretreated with Viscozyme.*

the intracellular essential oil and other bioactive compounds [22]. This observation could be noticed in **Figure 2**.

However, low solubility of essential oil in the aqueous buffer system restrains complete extraction [23, 24]. Therefore, this technique is used as a pretreatment to enhance extraction of volatile oil. The extraction of essential oils (EOs) is carried out by a single enzyme or a combination of enzymes; i.e. cellulase, hemicellulase, pectinase, protease, amylase or viscozyme (mixture of cellulase, arabinase, β-glucanase, hemicellulase and xylanase) [25]. Mahmoudi et al. [26] found that treating sweet basil (*Ocimum basilicum* L.) leaves with viscozyme before HD increased the EO yield to 9.32% instead of 6.1% in the control samples. Enzymatic pretreatment was carried out with 5 mg of enzyme/50 g leaves in distilled water (250 mL) during stirring at 40°C for 1 h prior to HD for 2 h. Li et al. [27] reported that EAE of EO from *Mentha haplocalyx* leaves with a mixture of enzymes (cellulase and pectinase) was more efficient than each of them. They attributed this effect to the synergistic influence of enzyme mixture on the cell wall. Moreover, the efficiency of EAE is affected by type and activity of enzymes used, enzyme concentration, buffer to solid ratio, incubation temperature, and incubation time [28, 29]. In addition, difference in plant structures affects the enzyme efficiency with respect to the oil yield [30].

Baby and Ranganathan [31, 32] pretreated cardamom and fennel seeds with either of celluclast, pectinex, viscozyme and protease prior to steam distillation led to an increase in EO yield by 7–16% and 11–22.5%, respectively. The maximum yield was obtained by viscozyme at 1% (*v/w*) under optimized conditions of 50°C, pH 5 and incubation time of 90 min. GC-MS analysis showed that enzyme pretreatment increased significantly the main characteristic oxygenated compounds (1,8-cineole and α-terpinyl acetate) in cardamom EO and *trans*-anethole and fenchone in fennel EO.

Increasing enzyme concentration enhances the degradation of cell wall and the release of EO up to a level after which no significant increase of EO yield could be obtained [32]. They ascribed this to the saturation of enzyme sites by the substrate. Shimotori et al. [33] treated peppermint (*Mentha arvensis*) leaves powder with enzyme aqueous solution at a ratio of 1:10 (*w/v*). Each of cellulase and hemicellulose was used at different concentrations (0.1, 1.0, 2.0, 5.0 and 10.0%, *w/w*), individually. The mixture was incubated for 3 h at 40°C, before subjecting to hydrodistillation for 1 h to obtain EO. Levels of L-menthol and L-menthone, the main components of EO, were used to examine the efficiency of the treatment. Maximum yields of L-menthol and L-menthone were obtained at each enzyme concentration of 2%. They found that the yield of EO increased by the combined use of 2% cellulase and 2% hemicellulase compared with the use of one enzyme.

Pretreatment of *Ocimum canum* aerial parts powder with viscozyme at 1% concentration before HD increased the EO yield to 1.2% compared to 0.83% with HD only [34]. This pretreatment decreased HD time from 180 to 30 min. They found that EAE increased the level of oxygenated monoterpenes in the obtained oil.

Vladić et al. [35] incubated *Origanum vulgare* aerial parts with viscozyme at 8%, pH 4.9, 45°C for 60 min before extraction for 4 h increased EO yield to 6.59% compared to 3.39% that obtained by the control. The EAE led to an increase of oxygenated compounds in the extracted EO to 94.67% compared to the control (88.06%).

### 2.3 Microwave-assisted extraction

Microwave-assisted extraction (MAE) accelerates the extraction of EO and saves energy and time without negative changes in the EO composition [36]. Microwaves (electromagnetic waves) rotate molecules with dipoles inside the plant cell that creates heat, generates high inwards pressure (as a result of the abrupt rise in temperature) on the cell wall disrupts it and releases cells' content into the extraction medium [37].

Hassanein et al. [38] extracted EO from the dried aerial parts of 7 plants from Lamiaceae family (*Origanum majorana* L., *Mentha pipereta* L., *Mentha longifolia* L., *Origanum syriacum* L., *Lavandula angustifolia* L., *Rosmarinus officinalis* L., and *Thymus vulgaris* L.) using MAHD at 100°C and 800 W for 60 min. The yield and oxygenated constituents of EO extracted by MAHD were higher than those of EO obtained by the HD technique in 180 min, indicating the higher quality of MAHD EOs. They reported that long extraction time by HD enhanced decomposition of the oxygenated compounds.

Ghazanfari et al. [39] extracted EO from coriander seeds powder by microwave-assisted hydrodistillation (MAHD). The microwave oven was operated as follows; 10 min at 800 W up to 100°C, and then kept at 100°C for 60 min at 500 W, followed by 10 min of ventilation. The EO yield (*v/w*) obtained by MAHD (0.325%) was not significantly different ($p > 0.05$) from that extracted by HD (0.31%). The MAHD technique reduced extraction time from 240 min during HD to 60 min.

Memarzadeh et al. [40] studied the effect of microwave-assisted steam hydrodiffusion (MSHD) technique and extraction time on the EO yield of the Bakhtiari savory (*S. bachtiarica* Bunge.) aerial parts. The MSHD was carried out at 800 W for 75 min. The highest EO yield (1.80, *v/w* dry weight basis) was obtained after 60 min of MSHD *vs* 150 min by the HD. The maximum level of oxygenated monoterpenes (69%) was obtained by MSHD after 20 min instead of 65.5% that extracted by the HD after 150 min. The MSHD technique reduced energy required for EO recovery from 4.5 kWh by the HD to 0.26 kWh.

Yingngam et al. [41] used solvent-free microwave extraction (SFME) attached to a Clevenger apparatus to retrieve EO from the fresh *Shorea roxburghii* flowers. The moisture content of the fresh flowers ranged between 69% and 74%. They subjected water inside plant cells to microwave energy. The effects of microwave power (480, 640, and 800 W) and irradiation time (10, 30, and 50 min) as independent variables, on the EO yield (%, *w/w*) were evaluated. The highest yield of EO (0.0114%) was recorded at 780 W and 38 min. The oil obtained by HD for 8 h had the similar yield. The EO recovered by SFME technique was characterized by the same scent of fresh flowers. The energy consumption decreased from 3.60 kWh by HD to 0.58 kWh with the SFME. In other research, Yingngam et al. [42] used the same SFME technique to extract EO from fresh aerial parts of *Limnophila aromatic* (70.11–75.14% moisture content). The highest yield of EO (0.21% *v/w*) was recovered at 700 W and 25 min instead of 4 h by HD. Oxygenated monoterpenes of the EO produced by SFME technique (50.29%) was higher than that obtained by HD (39.09%). However, they reported that the quality of the EO obtained by both methods was similar. Peng et al. [43] used a combined technique of solvent-free MAE (SFME) and the screw extrusion of *Pinus pumila* fresh needles to obtain the EO. Response surface method was applied to optimize the extraction conditions. The investigated independent factors used were; moisture content (30, 40, and 50%), MAE time (20, 30, and 40 min), and MAE power (385, 540, and 700 W) while yield of EO was the response factor. Fresh needles were crushed in the extrusion treatment and the cell wall was ruptured with the increase of intracellular pressure in the cells due to the evaporation of water (100°C) by the microwave irradiation. Increasing the moisture content of needles from 20% to 40%, increased the yield of EO. The highest yield (12.00 mL/kg) was obtained at moisture content of 40%, power of 540 W and irradiation time of 30 min. On the other hand, the EO yield from *P. pumila* did not exceed 7.00 mL/kg after 4 h of conventional HD.

### 2.4 Supercritical fluid extraction by $CO_2$

Supercritical fluid extraction (SFE) is a green, non-selective, and solvent free technique [44]. Carbon dioxide is applied in the SFE because it is cheap, non-flammable, has low critical temperature (31.1°C) and pressure (73.8 bar) and has a polarity appropriate for extraction of non-polar materials such as EO [45]. **Figure 3** illustrates a flow diagram of SFE.

This extraction technique, avoid degradation of thermolabile compounds compared to other techniques. However, the SFE has the disadvantage of high investment and operating costs [46].

Quintana et al. [47] reported that SFE produces high yield extracts with higher quality but lower concentration of volatile compounds compared with HD technique.

The efficiency of SFE process is affected by the following independent variables: particle size of the material, pressure, temperature, co-solvent and time [48].

The most appropriate particle size of the ground material is within the range of 0.4–0.8 mm [49]. The decrease of particle size (diameter) increases surface area of the material that is subjected to fluid $CO_2$ and enhances the extractability of the target components. Too small particle size reduces extract yield due to agglomeration of particles and reduction of surface area [50]. Under SFE conditions, increasing pressure at a specific temperature increases the $CO_2$ density, and consequently the solubility of the target compounds. However, increasing pressure above a certain level could result in a higher solubility of waxes and other hydrocarbons besides essential oil components. Meanwhile, increased temperature at constant pressure decrease the $CO_2$ density, and reduce the extraction yield though it increases the vapor pressure of

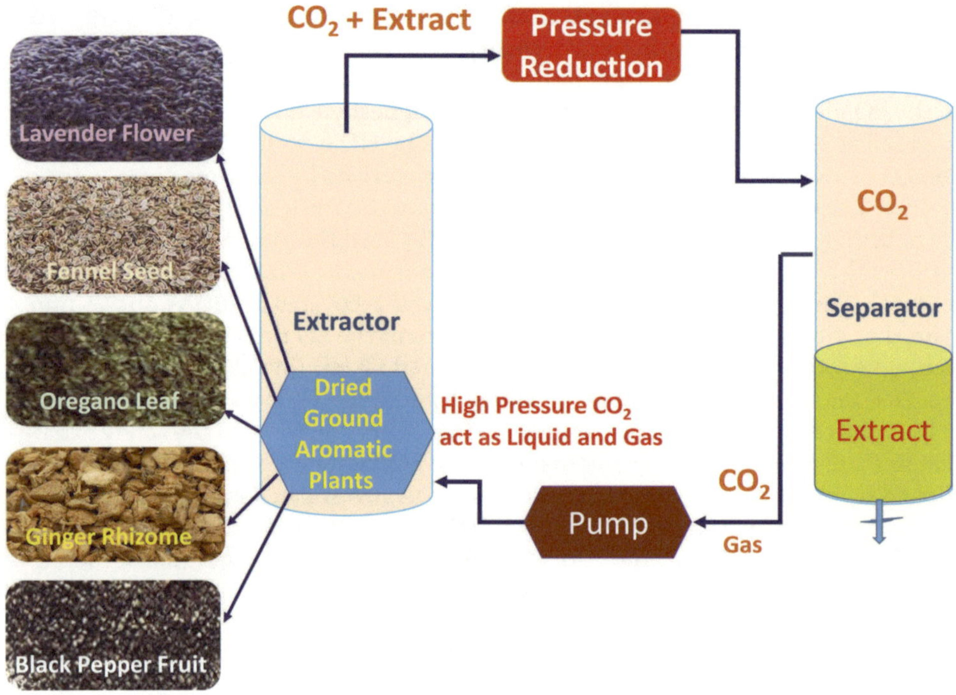


**Figure 3.**
*A flow diagram of supercritical fluid extraction by* $CO_2$.

the EO [51, 52]. The SFE temperature is used in the range 35–50°C to avoid degradation of thermolabile compounds. Since $CO_2$ is nonpolar, addition of small amounts of co-solvents (polar modifiers) such as methanol and ethanol increase the solubility of more polar compounds (phenolic compounds). The moisture content of the plant material can be used as a modifier. However, the modifier has to be separated from the resulted extract [45].

Markom et al. [53] studied the effect of co-solvents on the efficiency of SFE of EO from *Polygonum minus* dried leaves. The co-solvents used were: water, methanol, ethanol, and aqueous solutions of methanol and ethanol. The SFE was performed at 40°C and pressure of 150 bars. The $CO_2$ flow rate was 3 mL/min while the co-solvents flow rate was adjusted at 0.3 mL/min. The static and dynamic periods were set for 20 min and 240 min, respectively. The highest extraction yields (>25%) were obtained by the aqueous solutions of methanol and ethanol while the lowest yields (<9%) were obtained by the pure alcohols. The yield reached ~20% when water was used as a co-solvent.

Ara et al. [54] used central composite design to optimize the extraction yield of EO from *Descurainia sophia* L. ground dried seeds by SFE. The independent factors were: pressure (100, 228 and 355 bar), temperature (35, 50 and 65°C), modifier volume (methanol) (50, 100 and 150 μL), dynamic time (10, 25 and 40 min) and static extraction time (10, 25 and 40 min). The EO yield was used as a response. They found that increasing static extraction time from 10 to 40 min at the same extraction conditions (100 bar, 35°C, dynamic time of 10 min, without modifier) caused a slight increase in the yield from 0.5 to 1.1%. Increasing extraction temperature from 35 to 65°C at the same extraction conditions (100 bar, dynamic time of 40 min, static time of 10 min and modifier, 100 μL) resulted the same extraction yield (1.2%). Therefore, temperature and static time were fixed at 65°C and 10 min, respectively. Increasing pressure from 100 bar to 228 bar, at the same extraction conditions (methanol, 100 μL and dynamic time, 25 min), increased the extraction yield from 2.07 to 10.4%, due to increase of $CO_2$ density, which increases the solubility of the target components. Increasing modifier volume from 50 to 100 and 150 μL, at the same extraction conditions (228 bar and dynamic time, 25 min), increased the extraction yield from 9.24 to 10.4 and 12.72%, respectively. They attributed this increase to the ability of polar modifier to increase the solubility of polar compounds in the $CO_2$ and consequently, increases the extraction yield. The highest yield (18.48%) was obtained at the optimum conditions (355 bar, 65°C, static time of 10 min, dynamic time of 35 min and modifier volume of 150 μL).

Oliveira et al. [55] extracted EO fractions from *Piper divaricatum* dried leaves using SFE at temperature of 35°C and 55°C, and pressure of 100, 300 and 500 bar. Increasing pressure from 100 bar to 300 bar at 35°C caused an increase in the $CO_2$ density from 712.8 to 929.1 kg/m$^3$, which consequently increased significantly the EO yield from 4.68 to 6.03% dry weight basis. Further increase in pressure to 500 bar, at the same temperature, did not significantly increase the yield, though the $CO_2$ density increased to 1005 kg/m$^3$. The same trend was also noted when SFE was performed at 55°C using the same investigated levels of pressure. The yields at 55°C were found to be higher than those at 35°C. The highest yield (7.15%) was obtained at 55°C/300 bar instead of 3.03% that obtained after 3 h of HD. The highest concentrations of eugenol (21.7%) and methyl eugenol (61.85%) and the lowest concentration of eugenyl acetate (4.35%) were recorded for the EO obtained by HD. The concentration of eugenol did not exceed 13.2% in the extracts obtained by SFE. The SFE extracts were characterized by higher level of eugenyl acetate (>14.75%).

Marzlan et al. [56] extracted the EO from dried ground Torch ginger inflorescence with SFE at combinations of temperature of 34.7, 38, 46, 54, and 57.3°C and pressure of 83.6, 125, 225, 325 and 366.4 bar. The static and dynamic times were 2 h and 1 h, respectively. At a constant temperature (46°C), increasing pressure from 83.6 bar to 225 bar caused an increase in the EO yield from 0.65 to 5.2%, while continual increase in pressure to 366.4 bar resulted in a decrease in EO yield to 4.16%. On the other hand, at a constant pressure (225 bar), increasing temperature from 34.7 to 46°C caused an increase in the EO yield from 3.68 to 5.21%, while continual increase in temperature to 57.3°C resulted in a low increase in EO yield to 5.72%.

Silva et al. [57] obtained the EO from dried ground leaves of *Lippia thymoides* by SFE at 40 and 50°C, and pressures of 100, 200, and 300 bar. The static and the dynamic periods were 30 min and 120 min, respectively. The EO yield was 1.29% (*w/w*) at 200 bar and increased to ~1.6% (*w/w*) at 300 bar, regardless of the temperature used. The thymol (the major constituent, >74%) and oxygenated monoterpene contents of the EO obtained at 50°C were higher than those extracted at 40°C, regardless the level of pressure used.

## 3. Essential oil as a green preservative and flavoring agent

### 3.1 Essential oil as antioxidant agent

Oxidation of food products during processing and/or storage causes undesirable changes. It affects negatively nutritional quality and consumer acceptability (color changes and off-flavors). Antioxidants at low concentrations can delay oxidative reactions and extend the shelf life of the food products [58]. Many essential oils have antioxidant properties through scavenging of free-radicals and singlet oxygen quencher [59–62]. Essential oil constituents like thymol, eugenol, carvacrol, linalool, 1,8-cineole, geranial/neral, citronellal, isomenthone, and menthone are potent antioxidants. They can convert free radicals into more stable compounds by the addition of hydrogen atoms [63]. Strong antioxidant activity of essential oils is attributed to their phenolic structure [64].

The peroxide value (PV) is used for assessing the early stages of fat oxidation. Meanwhile, the thiobarbituric acid (TBA) value represented the secondary product (Malondialdehyde (MDA) of oxidation of polyunsaturated fatty acids. Direct addition of peppermint essential oil to refined soybean oil (without synthetic antioxidant) at 200 ppm or packaging it in a high density polyethylene package incorporated with 3700 ppm peppermint essential oil kept its oxidative stability not significantly different from that containing 200 ppm butylated hydroxyl toluene (BHT) during storage for 45 days at 40°C [65]. They attributed this activity to the main constituents: L-menthol, menthone and isomenthone of essential oil. Mezza et al. [66] found that addition of rosemary essential oil, obtained by hydrodistillation, to refined sunflower oil at 0.1 g/100 g extended its shelf life from 26 days to 36 days during storage in a dark place at 23°C, in presence of air. Meanwhile, the residue fractions of the essential oil, obtained by molecular distillation, displayed a longer shelf life that reached 44 days at the same storage conditions. The levels of less volatile constituents (camphor, α-terpineol and *cis*-sabinene hydrate) increased progressively with successive stages of molecular distillation. Okhli et al. [67] investigated the antioxidant properties of Citron peel essential oil, obtained by steam distillation, on sunflower oil at 800 ppm during storage at 65°C for 5 days. Oxidative stability of oil samples was measured with a Rancimat apparatus. Oxidative stability of oil samples enriched with essential oil (3.39 h) was higher than that of oil supplemented with 200 ppm BHT (3.0 h) at the end of storage period.

Immersion of Atlantic mackerel (*Scomber scombrus*) fillets in 1% (*w/v*) basil (*Ocimum basilicum*) and rosemary (*Rosmarinus officinalis*) essential oils for 30 min at 2°C and stored at the same temperature after packing into air-proofed polyamide/polyethylene packs delayed the development of lipid oxidation. TBA of the treated samples was followed during storage. TBA value exceeded the acceptable level (~5 mg MDA/kg of fish flesh according to Bensid et al. [68] after 8, 10, and 11 days for the control, rosemary, and basil groups, respectively. The basil and rosemary essential oils extended the shelf life of the fish fillets by 2 and 3 days, respectively, compared to the control group [69].

Boskovic et al. [70] evaluated the efficacy of thyme and oregano essential oils in retarding lipid oxidation of minced pork stored in modified conditions (vacuum and 30% $O_2$ conditions) at 3 ± 1°C during 15 days of storage. Minced pork samples were homogenized with different concentrations (0%, 0.3%, 0.6%, and 0.9%) of thyme or oregano essential oils. The control mince was prepared without essential oils. Essential oils reduced significantly ($p < 0.05$) the TBARS (mg malondialdehyde/kg) values in the mince even at the low level (0.3%). Minced samples with this concentration of essential oil were sensory acceptable. The antioxidant activity was attributed to the phenolic compounds in the investigated volatile oils.

Dipping raspberry fruits into lemon verbena essential oil emulsion at 250 μl/L for 3 min reduced the damage caused by reactive oxygen species during cold storage at 4°C for 9 days and extended shelf life of the fruits. The use of this essential oil as an edible coating increased antioxidant activity measured by inhibition of 2.2 diphenyl picrylhydrazyl (DPPH) radicals from 50.99 to 85.63% [71].

The DPPH radical-scavenging activity of edible films based on konjac glucomannan (KGM) polysaccharide loaded with thyme essential oil (TEO) at 0.4, 0.8, 1.2 and 1.6% (*v/v*) increased significantly ($p < 0.05$) with the increase of TEO concentration. Loading KGM-based films with TEO at 1.6% increased its antioxidant capacity by about 50% [72]. They suggested using these films loaded with TEO in food packaging.

Food grade nano-emulsions have been widely used to enhance the water solubility and stability of essential oils [73]. The cellulose nanofibrils films prepared by O/W Pickering emulsion with oregano essential oil exhibit excellent antioxidant activity [74]. They attributed this activity to phenolic compounds of oregano essential oil and recommended these films for packaging of easily oxidized foods.

Cinnamon and clove essential oils improved the antioxidant capacity of mandarin (*Citrus reticulata*) essential oil nanocapsules by 4.43 and 3.52 times, respectively. This increment of antioxidant activity was attributed to the high antioxidant activity of cinnamon and clove essential oils components [75].

Nanoemulsion-based basil seed gum (NBSG) films containing clove essential oil (CEO) had higher antioxidant activity than that of BSG films prepared by conventional method with the same concentration of CEO. The DPPH and ABTS radical scavenging activities of NBSG-CEO films containing 10 mg CEO/mL were not significantly different from those of NBSG-BHT films containing 1 mg BHT/mL. Eugenol is the main constituent of CEO. Wrapping minced camel meat sample with NBSG films containing resveratrol (4 μg/mL) + CEO (10 mg/mL) kept its oxidative stability after 20 days of storage at 4°C better than that of the control group. Peroxide and thiobarbituric acid (TBA) values of the NBSG wrapped meat samples did not exceed 4.03 meq/kg lipid and 1.03 mg malondialdehyde/kg after 20 days of cold storage [76].

Kiralan et al. [77] flavored olive oil with peppermint, oregano, thyme and laurel essential oils at 0.05% (*v/w*). GC-MS analysis of the headspace of flavored samples was carried out after 15, 30, and 45 days of storage at 60°C. The major components of essential oil transferred into olive oil samples. *E*-2-hexenal and hexanal were the main volatile constituents of olive oils during oxidation. After 30 days of thermal oxidation the *E*-2-hexenal level of the control and the peppermint flavored oil exceeded its initial level by 30 and 90 times, respectively. Flavoring olive oil with oregano, thyme and laurel essential oils kept *E*-2-hexenal level in the headspace of flavored samples during thermal oxidation lower than 2 times its original level.

### 3.2 Essential oil as antimicrobial agent

EOs with antimicrobial activity inhibit the microbial cells reproductive ability, or damage bacterial cells [78, 79]. Hydrophobicity/lipophilicity property of EOs allows them to cross the cell cytoplasmic membrane and raising permeability of fatty acids, polysaccharides, and phospholipid layers [80]. This causes leakage of cell contents, and loss of macromolecules.

EOs have the ability to coagulate the cytoplasm and inhibit enzymes responsible for the synthesis of biologically active components [81]. Gram positive-bacteria are more sensitive to EOs effect than gram-negative ones, since the outer membrane surrounding the cell wall of gram negative-bacteria, restricts the penetration of EOs through the lipopolysaccharide layer [62].

EOs rich in phenolic compounds like thymol, carvacrol or eugenol display high antimicrobial activities against foodborne pathogens [82, 83]. Aromatic plants that are rich in these phenolic compounds are illustrated in **Figure 4**.

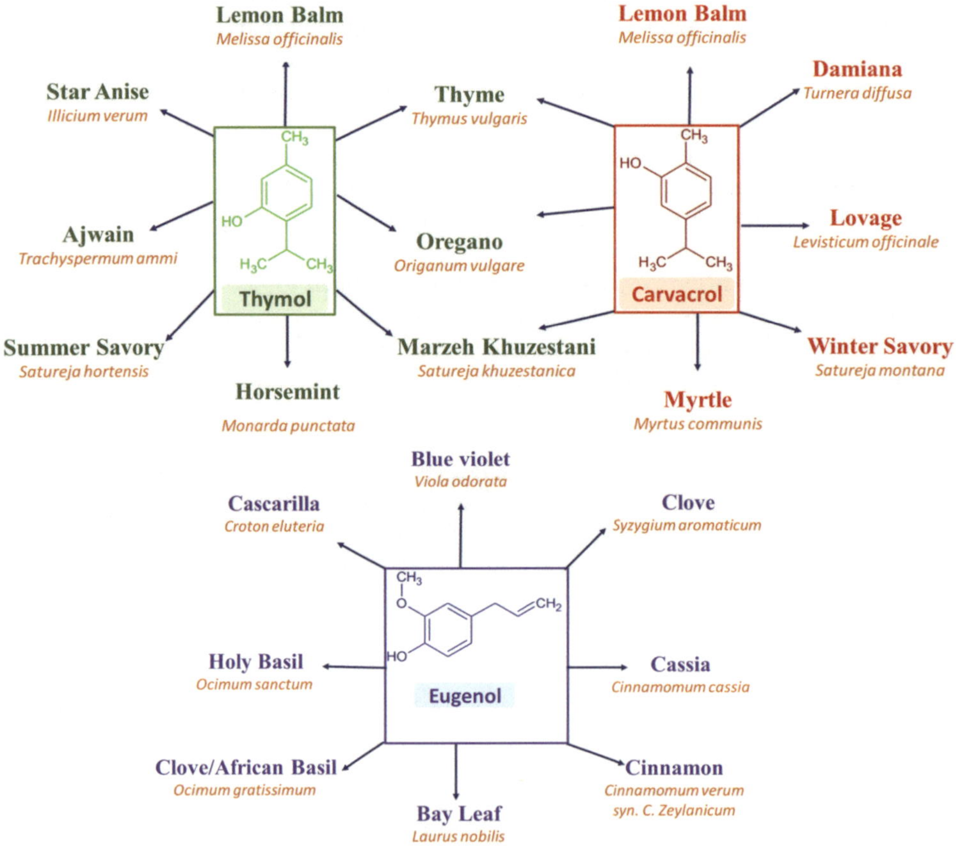


**Figure 4.**
*Examples of some aromatic plants rich in thymol, carvacrol and eugenol as phenolic components found in essential oils of these plants.*

These phenolic compounds attack the amine groups in the cell membrane, alter its permeability leading to cell lysis [7, 84]. Moreover, a synergistic effect between EOs components enhances its antimicrobial efficiency. The synergistic effect between ρ-cymene and carvacrol, geraniol and menthol is a good example against wide bacteria range (**Figure 5**).

This antimicrobial efficiency was significantly weaker when each compound acted separately in the same medium. Some EO constituents do not possess antibacterial properties when used alone, but they enhance the bacterial inhibition of other compounds [85].

The incorporation of lemongrass (*Cymbopogon citratus*) essential oil (LEO) into chitosan-based films at 9% level controlled the growth of Gram-positive bacteria (*B. cereus* and *L. monocytogenes*) and Gram-negative bacteria (*E. coli* and *S. typhi*). The antibacterial activity of the films was evaluated using the disk-diffusion method. The chitosan/LEO composite film with 9% LEO completely inhibited the growth of *S. typhi* [86].

Chitosan, gum arabic, and polyethylene glycol composite film incorporated with black pepper essential oil or ginger essential oil possessed high antimicrobial activity against *Bacillus cereus*, *Staphylococcus aureus, Escherichia coli,* and *Salmonella typhimurium* [87].

*DOI: http://dx.doi.org/10.5772/intechopen.103035*

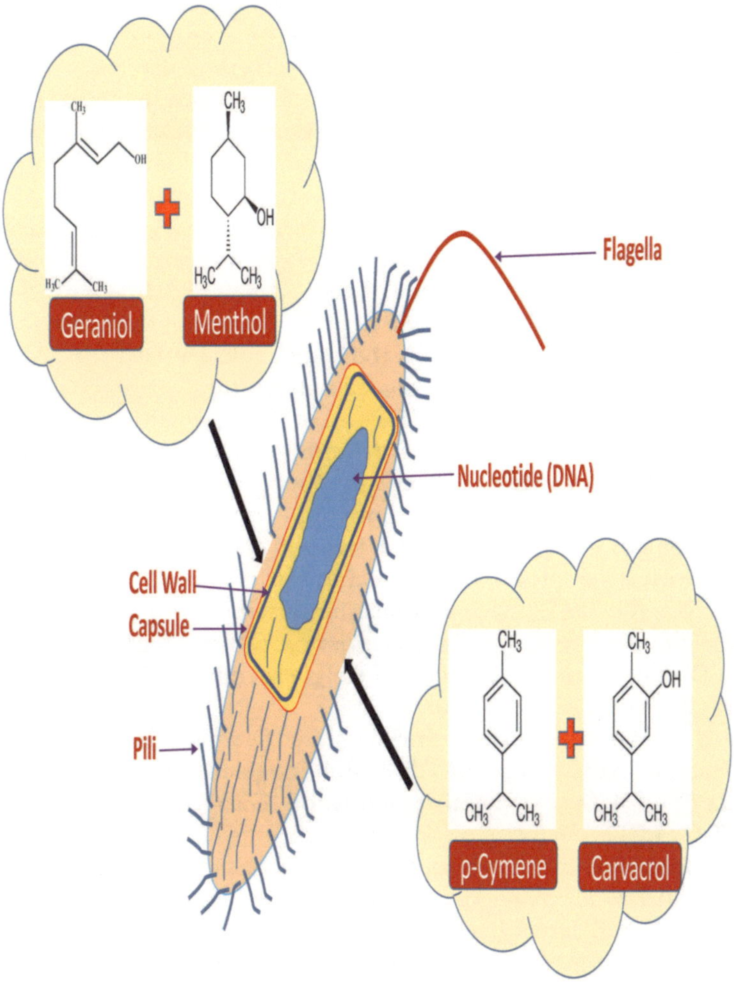


**Figure 5.**
*Schematic representation the synergistic effect of ρ-cymene and carvacrol, geraniol and menthol mechanism of action as antimicrobial.*

Fattahian et al. [88] investigated he antimicrobial activity of cumin essential oil (EO) against *Staphylococcus aureus* and *Escherichia coli* O157:H7. They found that *Staphylococcus aureus* was more sensitive to EO than *E. coli* O157:H7. Meanwhile, they studied the effect of coating fresh veal fillets with a biodegradable film of chitosan (CH) incorporated with cumin nanoliposomal EO at 1% level on the microbial properties of veal samples stored in modified atmosphere packages (20% $CO_2$ and 80% $O_2$) at 4°C for 21 days. The total microbial count, lactic acid bacteria, enterobacteriaceae, and pseudomonas were used as microbiological indicators. Encapsulation of cumin EO controlled the release of the antimicrobial compounds on coated samples that extended antimicrobial activity during cold storage for 21 days compared to free EO. At the end of storage, the investigated bacterial strains count of the CH + Nano EO coated samples were lower than those of CH + EO groups. Both coating films kept the bacterial load of meat fillets less than the CH group till the end of storage time.

The antibacterial activity was attributed to cuminaldehyde and the phenolic compounds in cumin essential oil and the synergistic effect of chitosan with EO.

Langroodi et al. [89] coated turkey breast fillet with chitosan incorporated with 1% (*v/v*) of *Origanum vulgare* essential oil and dried ethanolic extract of grape seeds (GSE, 2% *v/v*) before storage at 4°C for 20 days. Alterations in total viable count (TVC), lactic acid bacteria (LAB), Enterobacteriaceae, *Pseudomonas* spp., and yeast-mold counts of cold stored turkey meat samples and sensorial properties of roasted (10 min at 100°C) turkey samples were studied. Incorporation of oregano EO and GSE into chitosan increased the antibacterial activity of the coating film. The TVC counts of control and chitosan coated samples turned unacceptable (>6 CFU/g) after 12 and 16 days of storage, respectively. Coating with chitosan containing oregano essential oil and GSE kept TVC count of the samples less than 5 Log CFU/g after 20 days of storage. At the end of cold storage, the LAB and Enterobacteriaceae counts of samples coated with chitosan-oregano essential oil- and GSE were ~ 4 and 4.39 Log CFU/g instead of 7.22 and 7.14 Log CFU/g for the chitosan coated samples, respectively. Furthermore, inclusion of oregano essential oil and GSE into chitosan coating reduced the Pseudomonads counts in the samples by ~3 Log CFU/g, at the end of storage time. Application of oregano EO at 1% level and GSE at 2% enhanced the antifungal activity of chitosan coating. Yeast-mold count of turkey meat samples coated with chitosan-oregano EO-GSE did not exceed 4.27 Log CFU/g at the end of storage time. They attributed this strong antibacterial activity of the coating film to the synergistic effect of oregano EOs and GSE. CH-GSE 2%-O coated samples depicted the highest consumer scores compared with other samples till the end of storage.

Sayadi et al. [90] packaged fresh chicken pieces (2 cm thickness and 20 cm length) in plain and nano composite edible films of gelatin (GE) containing 1% $TiO_2$ nanoparticles (GE + $TiO_2$), 2% cumin essential oil (GE + CEO), and 1% $TiO_2$ + 2% CEO (GE + $TiO_2$ + CEO) before storing in polyethylene plastic bags at 4 ± 1°C for 24 days. The population of total mesophilic bacteria, Enterobacteriaceae, lactic acid bacteria, and *Pseudomonas* spp., of packaged samples were evaluated. The bacterial growth (for different bacteria) in the control increased by ≥4 log CFU/g, after 24 days of storage. At the end of storage time the lowest population (<6 log CFU/g) of the tested bacteria was recorded for GE + CEO and GE + $TiO_2$ + CEO chicken samples. They attributed the antimicrobial activity to cuminaldehyde component and phenolic compounds of CEO in addition to reactive oxygen species generated by $TiO_2$-N that disrupt the bacteria membrane. Although the GE + CEO and GE + $TiO_2$ + CEO groups obtained the highest sensory scores among chicken samples, both groups showed unacceptable scores of sensory attributes (odor and overall acceptability) after 16 days of storage.

Sharma et al. [91] tested essential oils of clove bud, tagetes, thyme, eucalyptus, neem, cinnamon leaf, himalayan pine needle, and tea tree against the total bread molds by agar well-diffusion method. Thyme oil completely inhibited the growth of bread molds than other essential oils. They found that sealing fresh white slices of bread in a biodegradable film (poly, 3-hydroxybutyrate-co-4-hydroxybutyrate) incorporated with thyme essential oil at 30% (*v/w*) extended its shelf-life against molds to >5 days at ambient conditions (25–28°C and 35–45% RH), compared to 1–4 days in neat biopolymer film. The molds count of the bread packaged in this film after 5 days of storage was <1.00 log (CFU/mL), the same level at the zero-time storage.

### 3.3 Essential oil as flavoring agent

Flavoring is one of the main application of essential oils in the food and beverage industries [81]. Flavorings are used to improve the odor of foods in order to satisfy the consumer. EOs are used in the preparation of carbonated beverage to give the product its distinctive aroma.

Recently, flavored edible oils are produced in order to improve its sensory properties [92, 93]. Flavoring of oils is carried out by infusion or maceration of the aromatic plant into the oil [94, 95]. Theses flavoring techniques enhance the extraction of waxes and undesirable components into oil [96]. To overcome these defects, EOs have been used as flavoring materials [97]. However, strong flavors with some EOs may negatively affect the consumer acceptability of the food. The flavoring of edible oils improves their sensory properties [92], increase their use for the preparation of daily food condiments [98] and extend their usage by non-traditional consumers [99].

Porto and Decorti [100] flavored ricotta cheese with thyme essential oil by mixing at 0.26, 0.33 and 0.40% (*w/w*). The main constituents of the essential oil were carvacrol, carvacrol methyl ether, ρ-cymene, γ-terpinene and thymol. Sensory studies indicated that the minimum perception level of thyme essential oil in ricotta cheese was 0.20% (*w/w*). Aroma compounds of the flavored ricotta cheese were extracted by Headspace solid-phase micro-extraction at 30°C and were identified with GC-MS. Results showed that hydrocarbons monoterpene and hydrocarbons sesquiterpene were lower in the headspace of ricotta by 25% and 40%, respectively, compared to their original level in essential oil. They attributed these decrements to the binding capacity of fat and proteins to flavor compounds.

Benkhoud et al. [101] flavored extra virgin olive oil by homogenization with Eos (500 ppm) of thyme, rosemary, black pepper, fennel, and citrus peels. Flavored oil samples were stored for 12 months, at room temperature. The headspace of flavored oil enriched mainly with the major components of the flavoring essential oils. They attributed the bitterness of rosemary, thyme, and fennel flavored samples to the presence of 1,8-cineole and carvacrol while pungency of black pepper flavored samples was ascribed to β-caryophyllene and α-phellandrene. Citrus-flavored oil samples were distinguished by their fruity taste due to limonene. The highest acceptability scores were recorded for fennel and citrus flavored oil samples.

Moustakime et al. [99] flavored virgin olive oil (VOO) with the seeds of green anise. The main component of the anise seeds EO was found to be *trans*-anethole (76.16%). This compound was used as an indicator for the level of flavoring. Flavoring of VOO was carried out with anise seeds at a ratio of 15% (w/w) with maceration, sonication (intensity ~1 $W/cm^2$) and direct addition of the EO (0.33 mL, equivalent to amount of oil from 15 g seeds) using stirring for 24 h. GC/MS analysis indicated that the diffusible *trans*-anethole reached 26.59% of the total volatile fraction of the flavored oil after 15 min of ultrasound treatment instead of 23.85% after 9 days of maceration. Meanwhile, *trans*-anethole level of the total volatile fraction reached 36.3% by direct addition of EO.

## 4. Conclusions

The ecofriendly techniques meet the terms of green extraction, reduces extraction time, with high yield, low energy consumption and solvent amount, allows the use of renewable natural products, and ensures a safe and high-quality essential oil. The addition of essential oils to food as green preservative causes many positive effects such as antioxidant, antimicrobial activities and improve the flavor in food products. This effect could be due to the synergistic combination of the essential oil constituents rather than one component.

## Conflict of interest

"The authors declare no conflict of interest."

## Thanks

I would like to express my special thanks and gratitude to Professor Samy Mohamed Galal, Food Science Department, Faculty of Agriculture, Cairo University, who encouraged me to write this chapter, and he also helped me in revising it. I appreciated his efforts.

## Author details

Nashwa Fathy Sayed Morsy
Faculty of Agriculture, Food Science Department, Cairo University, Giza, Egypt

*Address all correspondence to: nanafsm@yahoo.com

## References

[1] Liu XY, Ou H, Xiang ZB, Gregersen H. Optimization, chemical constituents and bioactivity of essential oil from *Iberis amara* seeds extracted by ultrasound assisted hydro-distillation compared to conventional techniques. Journal of Applied Research on Medicinal and Aromatic Plants. 2019;**13**:100204. DOI: 10.1016/j.jarmap.2019.100204

[2] Ferhat MA, Meklati BY, Smadja J, Chemat F. An improved microwave Clevenger apparatus for distillation of essential oils from orange peel. Journal of Chromatography A. 2006;**1112**:121-126. DOI: 10.1016/j.chroma.2005.12.030

[3] Mohamad N, Ramli N, Abd-Aziz S, Ibrahim MF. Comparison of hydro-distillation, hydro-distillation with enzyme-assisted and supercritical fluid for the extraction of essential oil from pineapple peels. 3 Biotech. 2019;**9**:234. DOI: 10.1007/s13205-019-1767-8

[4] Huang P, Gu Z, Yang L, Yang R, Ji Y, Zeng Q, et al. Study on optimization of extraction technique of pericarp essential oil in *Litsea Cubeba* (Lour) Pers. Journal of Food Measurement and Characterization. 2021;**15**:758-768. DOI: 10.1007/s11694-020-00657-0

[5] Fernández-López J, Viuda-Martos M. Introduction to the special issue: Application of essential oils in food systems. Foods. 2018;**7**(4):56. DOI: 10.3390/foods7040056

[6] Pelaes Vital AC, Guerrero A, Guarnido P, Cordeiro Severino I, Olleta JL, Blasco M, et al. Effect of active-edible coating and essential oils on lamb patties oxidation during display. Foods. 2021;**10**(2):263. DOI: 10.3390/foods10020263

[7] Hyldgaard M, Mygind T, Meyer RL. Essential oils in food preservation: Mode of action, synergies, and interactions with food matrix components. Frontiers in Microbiology. 2012;**3**:1-24. DOI: 10.3389/fmicb.2012. 00012

[8] Simionato I, Domingues FC, Nerín C, Silva F. Encapsulation of cinnamon oil in cyclodextrin nanosponges and their potential use for antimicrobial food packaging. Food and Chemical Toxicology. 2019;**132**:110647. DOI: 10.1016/j.fct.2019.110647

[9] Chaudhari AK, Dwivedy AK, Singh VK, Das S, Singh A, Dubey NK. Essential oils and their bioactive compounds as green preservatives against fungal and mycotoxin contamination of food commodities with special reference to their nanoencapsulation. Environmental Science and Pollution Research. 2019;**26**:25414-25431. DOI: 10.1007/s11356-019-05932-2

[10] Rao MV, Sengar AS, Sunil CK, Rawson A. Ultrasonication—A green technology extraction technique for spices: A review. Trends in Food Science and Technology. 2021;**116**:975-991. DOI: 10.1016/j.tifs.2021.09.006

[11] Muhammad DRA, Tuenter E, Patria GD, Foubert K, Pieters L, Dewettinck K. Phytochemical composition and antioxidant activity of *Cinnamomum burmannii* Blume extracts and their potential application in white chocolate. Food Chemistry. 2021;**340**:127983. DOI: 10.1016/j.foodchem.2020.127983

[12] Arya P, Kumar P. Comparison of ultrasound and microwave assisted extraction of diosgenin from *Trigonella foenum graceum* seed. Ultrasonics Sonochemistry. 2021;**74**:105572. DOI: 10.1016/j.ultsonch.2021.105572

[13] Chemat F, Rombaut N, Sicaire AG, Meullemiestre A, Fabiano-Tixier AS, Abert-Vian M. Ultrasound assisted extraction of food and natural products. Mechanisms, techniques, combinations, protocols and applications. A review. Ultrasonics Sonochemistry. 2017;**34**:540-560. DOI: 10.1016/j.ultsonch.2016.06.035

[14] Chen G, Sun F, Wang S, Wang W. Enhanced extraction of essential oil from *Cinnamomum cassia* bark by ultrasound assisted hydrodistillation. Chinese Journal of Chemical Engineering. 2020;**36**:38-46. DOI: 10.1016/j.cjche.2020.08.007

[15] Lilia C, Abdelkader A, Karima AKA, Tarek B. The effect of ultrasound pre-treatment on the yield, chemical composition and antioxidant activity of essential oil from Wild *Lavandula stoechas* L. Journal of Essential Oil Bearing Plants. 2018;**21**:253-263. DOI: 10.1080/0972060X.2018.1432419

[16] Wu Y, Jiang X, Zhang L, Zhou Y. Ultrasonic-assisted extraction, comparative chemical composition and biological activities of essential oils of fresh and dry aboveground parts of *Artemisia annua* L. Journal of Essential Oil Bearing Plants. 2018;**21**:1624-1635. DOI: 10.1080/0972060X.2019.1574244

[17] Jadhav NL, Garule PA, Pinjari DV. Comparative study of ultrasound pretreatment method with conventional hydrodistillation method for extraction of essential oil from *Piper betle* L. (Paan). Indian Chemical Engineer. 2020. DOI: 10.1080/00194506.2020.1828193

[18] Ghule SN, Desai MA. Intensified extraction of valuable compounds from clove buds using ultrasound assisted hydrotropic extraction. Journal of Applied Research on Medicinal and Aromatic Plants. 2021;**25**:100325. DOI: 10.1016/j.jarmap.2021.100325

[19] Guo J, Yang R, Gong Y, Hu K, Hu Y, Song F. Optimization and evaluation of the ultrasound-enhanced subcritical water extraction of cinnamon bark oil. LWT—Food Science and Technology. 2021;**147**:111673. DOI: 10.1016/j.lwt.2021.111673

[20] Zhang X, Zhu H, Wang J, Li F, Wang J, Ma X, et al. Anti-microbial activity of citronella (*Cymbopogon citratus*) essential oil separation by ultrasound assisted ohmic heating hydrodistillation. Industrial Crops and Products. 2022;**176**:114299. DOI: 10.1016/j.indcrop.2021.114299

[21] Reis NS, Brito AR, Pacheco CSV, Costa LCB, Gross E, Santos TP, et al. Improvement in menthol extraction of fresh leaves of *Mentha arvensis* by the application of multi-enzymatic extract of *Aspergillus niger*. Chemical Engineering Communications. 2019;**206**:387-397. DOI: 10.1080/00986445.2018.1494580

[22] Liu Z, Li H, Cui G, Wei M, Zou Z, Ni H. Efficient extraction of essential oil from *Cinnamomum burmannii* leaves using enzymolysis pretreatment and followed by microwave-assisted method. LWT—Food Science and Technology. 2021;**147**:111497. DOI: 10.1016/j.lwt.2021.111497

[23] Gligor O, Mocan A, Moldovan C, Locatelli M, Crișan G, Ferreira ICFR. Enzyme-assisted extractions of polyphenols: A comprehensive review. Trends in Food Science and Technology. 2019;**88**:302-315. DOI: 10.1016/j.tifs.2019.03.029

[24] Patil PD, Patil SP, Kelkar RK, Patil NP, Pise PV, Nadar SS. Enzyme-assisted supercritical fluid extraction: An integral approach to extract bioactive compounds. Trends in Food Science and Technology. 2021;**116**:357-369. DOI: 10.1016/j.tifs.2021.07.032

[25] Chiang CF, Lai LS. Effect of enzyme-assisted extraction on the physicochemical properties of mucilage from the fronds of *Asplenium australasicum* (J. Sm.) Hook. International Journal of Biological Macromolecules. 2019;**124**:346-353. DOI: 10.1016/j.ijbiomac.2018.11.181

[26] Mahmoudi H, Marzouki M, M'Rabet Y, Mezni M, Ait Ouazzou A, Hosni K. Enzyme pretreatment improves the recovery of bioactive phytochemicals from sweet basil (*Ocimum basilicum* L.) leaves and their hydrodistilled residue by-products, and potentiates their biological activities. Arabian Journal of Chemistry. 2020;**13**:6451-6460. DOI: 10.1016/j.arabjc.2020.06.003

[27] Li Z, Wang H, Pan X, Guo Y, Gao W, Wang J, et al. Enzyme-deep eutectic solvent pre-treatment for extraction of essential oil from *Mentha haplocalyx* Briq. leaves: Kinetic, chemical composition and inhibitory enzyme activity. Industrial Crops and Products. 2022;**177**:114429. DOI: 10.1016/j.indcrop.2021.114429

[28] Pontillo ARN, Papakosta-Tsigkri L, Lymperopoulou T, Mamma D, Kekos D, Detsi A. Conventional and enzyme-assisted extraction of rosemary leaves (*Rosmarinus officinalis* L.): Toward a greener approach to high added-value extracts. Applied Sciences. 2021;**11**:3724. DOI: 10.3390/app11083724

[29] Picot-Allain C, Mahomoodally MF, Ak G, Zengin G. Conventional versus green extraction techniques: A comparative perspective. Current Opinion in Food Science. 2021;**40**: 144-156. DOI: 10.1016/j.cofs.2021.02.009

[30] Hosni K, Hassen I, Chaabane H, Jemli M, Dallali S. Enzyme-assisted extraction of essential oils from thyme (*Thymus capitatus* L.) and rosemary (*Rosmarinus officinalis* L.): Impact on yield, chemical composition and antimicrobial activity. Industrial Crops and Products. 2013;**47**:291-299. DOI: 10.1016/j.indcrop.2013.03.023

[31] Baby KC, Ranganathan TV. Effect of enzyme pre-treatment on extraction yield and quality ofcardamom (*Elettaria cardamomum* maton.) volatile oil. Industrial Crops and Products. 2016a;**89**:200-206. DOI: 10.1016/j.indcrop.2016.05.017

[32] Baby KC, Ranganathan TV. Effect of enzyme pre-treatment on extraction yield and quality of fennel (*Foeniculum vulgare*) volatile oil. Biocatalysis and Agricultural Biotechnology. 2016b;**8**: 248-256. DOI: 10.1016/j.bcab.2016.10.001

[33] Shimotori Y, Watanabe T, Kohari Y, Chiou T, Ohtsu N, Nagata Y, et al. Enzyme-assisted extraction of bioactive phytochemicals from Japanese peppermint (*Mentha arvensis* L. cv. 'Hokuto'). Journal of Oleo Science. 2020;**69**(6):635-642. DOI: 10.5650/jos.ess19181

[34] Morsy NFS, Hammad KSM. Extraction of essential oil from methyl cinnamate basil (*Ocimum canum* Sims) with high yield in a short time using enzyme pretreatment. Journal of Science and Technology. 2021;**58**(7):2599-2605. DOI: 10.1007/s13197-020-04766-y

[35] Vladić J, Duarte ARC, Radman S, Simić S, Jerković I. Enzymatic and microwave pretreatments and supercritical $CO_2$ extraction for improving extraction efficiency and quality of *Origanum vulgare* L. spp. hirtum extracts. Plants. 2022;**11**:54. DOI: 10.3390/plants11010054

[36] Memarzadeh SM, Pirbalouti AG, AdibNejad M. Chemical composition and yield of essential oils from Bakhtiari

savory (*Satureja bachtiarica* Bunge.) under different extraction methods. Industrial Crops and Products. 2015;**76**:809-816. DOI: 10.1016/j.indcrop.2015.07.068

[37] Kazemi M, Niazi A, Yazdanipour A. Extraction of *Satureja rechingeri* volatile components through ultrasound-assisted and microwave-assisted extractions and comparison of the chemical composition with headspace solid-phase microextraction. Journal of Essential Oil Research. 2022;**34**. DOI: 10.1080/10412905.2021.1975575

[38] Hassanein HD, El-Gendy AE-NG, Saleh IA, Hendawy SF, Elmissiry MM, Omer EA. Profiling of essential oil chemical composition of some Lamiaceae species extracted using conventional and microwave-assisted hydrodistillation extraction methods via chemometrics tools. Flavour and Fragrance Journal. 2020;**35**:329-340. DOI: 10.1002/ffj.3566

[39] Ghazanfari N, Mortazavi SA, Yazdi FT, Mohammadi M. Microwave-assisted hydrodistillation extraction of essential oil from coriander seeds and evaluation of their composition, antioxidant and antimicrobial activity. Heliyon. 2020;**6**:e04893. DOI: 10.1016/j.heliyon.2020.e04893

[40] Memarzadeh SM, Gholami A, Pirbalouti AG, Masoum S. Bakhtiari savory (*Satureja bachtiarica* Bunge.) essential oil and its chemical profile, antioxidant activities, and leaf micromorphology under green and conventional extraction techniques. Industrial Crops and Products. 2020;**154**:112719. DOI: 10.1016/j.indcrop.2020.112719

[41] Yingngam B, Navabhatra A, Brantner A. Increasing the essential oil yield from *Shorea roxburghii* inflorescences using an eco-friendly solvent-free microwave extraction method for fragrance applications. Journal of Applied Research on Medicinal and Aromatic Plants. 2021;**24**:100332. DOI: 10.1016/j.jarmap.2021.100332

[42] Yingngam B, Brantner A, Treichler M, Brugger N, Navabhatra A, Nakonrat P. Optimization of the eco-friendly solvent-free microwave extraction of *Limnophila aromatica* essential oil. Industrial Crops and Products. 2021;**165**:113443. DOI: 10.1016/j.indcrop.2021.113443

[43] Peng X, Yang X, Gu H, Yang L, Gao H. Essential oil extraction from fresh needles of *Pinus pumila* (Pall.) Regel using a solvent-free microwave-assisted methodology and an evaluation of acetylcholinesterase inhibition activity in vitro compared to that of its main components. Industrial Crops and Products. 2021;**167**:113549. DOI: 10.1016/j.indcrop.2021.113549

[44] Khaw KY, Parat MO, Shaw PN, Falconer JR. Solvent supercritical fluid technologies to extract bioactive compounds from natural sources: A review. Molecules. 2017;**22**:1186. DOI: 10.3390/molecules22071186

[45] Yousefi M, Rahimi-Nasrabadi M, Pourmortazavi SM, Wysokowski M, Jesionowski T, Ehrlich H, et al. Supercritical fluid extraction of essential oils. TrAC Trends in Analytical Chemistry. 2019;**118**:182-193. DOI: 10.1016/j. trac.2019.05.038

[46] Cvjetko M, Jokić S, Lepojević Ž, Vidović S, Marić B, Redovniković IR. Optimization of the supercritical $CO_2$ extraction of oil from rapeseed using response surface methodology. Food Technology and Biotechnology. 2012;**50**:208-215

[47] Somaris E, Quintana SE, Llalla O, Garcia-Risco MR, Fornari T. Comparison

between essential oils and supercritical extracts into chitosan based edible coatings on strawberry quality during cold storage. The Journal of Supercritical Fluids. 2021;**171**:105198. DOI: 10.1016/j.supflu.2021.105198

[48] Fornari T, Vicente G, Vázquez E, García-Risco MR, Reglero G. Isolation of essential oil from different plants and herbs by supercritical fluid extraction. Journal of Chromatography A. 2012;**1250**:34-48. DOI: 10.1016/j.chroma.2012.04.051

[49] Bubalo CM, Vidović S, Redovniković IR, Jokić S. New perspective in extraction of plant biologically active compounds by green solvents. Food and Bioproducts Processing. 2018;**109**:52-73. DOI: 10.1016/j.fbp.2018.03.001

[50] Xiong K, Chen Y. Supercritical carbon dioxide extraction of essential oil from tangerine peel: Experimental optimization and kinetics modelling. Chemical Engineering Research and Design. 2020;**164**:412-423. DOI: 10.1016/j.cherd.2020.09.032

[51] Uwineza PA, Waskiewicz A. Recent advances in supercritical fluid extraction of natural bioactive compounds from natural plant materials. Molecules. 2020;**25**:3847. DOI: 10.3390/molecules25173847

[52] Leila M, Ratiba D, Al-Marzouqi A-H. Experimental and mathematical modelling data of green process of essential oil extraction: Supercritical $CO_2$ extraction. Materials Today: Proceedings. 2022;**49**:1023-1029. DOI: 10.1016/j.matpr.2021.08.125

[53] Markom M, Hassim N, Anuar N, Baharum S. Co-solvent selection for supercritical fluid extraction of essential oil and bioactive compounds from *Polygonum minus*. ASEAN Journal of Chemical Engineering. 2013;**12**:19-26. DOI: 10.22146/ajche.49739

[54] Ara KM, Jowkarderis M, Raofie F. Optimization of supercritical fluid extraction of essential oils and fatty acids from flixweed (*Descurainia sophia* L.) seed using response surface methodology and central composite design. Journal of Food Science and Technology. 2015;**52**:4450-4458. DOI: 10.1007/s13197-014-1353-3

[55] Oliveira MS, da Cruz JN, Silva SG, da Costa WA, de Sousa SHB, Bezerra FWF, et al. Phytochemical profile, antioxidant activity, inhibition of acetylcholinesterase and interaction mechanism of the major components of the *Piper divaricattun* essential oil obtained by supercritical $CO_2$. The Journal of Supercritical Fluids. 2019;**145**:74-84. DOI: 10.1016/j.supflu.2018.12.003

[56] Marzlan AA, Muhialdin BJ, AbedinNHZ,MohammedNK,AbadlMMT, Roby BHM, et al. Optimized supercritical $CO_2$ extraction conditions on yield and quality of Torch Ginger (*Etlingera Elatior* (Jack) RM Smith) inflorescence essential oil. Industrial Crops and Products. 2020;**154**:112581. DOI: 10.1016/j.indcrop.2020.112581

[57] Silva SG, de Oliveira MS, Cruz JN. Supercritical $CO_2$ extraction to obtain *Lippia thymoides* Mart. & Schauer (Verbenaceae) essential oil rich in thymol and evaluation of its antimicrobial activity. The Journal of Supercritical Fluids. 2021;**168**:105064. DOI: 10.1016/j.supflu.2020.105064

[58] Amorati R, Foti MC, Valgimigli L. Antioxidant activity of essential oils. Journal of Agricultural and Food Chemistry. 2013;**61**(46):10835-10847. DOI: 10.1021/jf403496k

[59] Maqsood S, Benjakul S, Abushelaibi A, Alam A. Phenolic compounds and plant phenolic extracts as natural antioxidants in prevention of lipid oxidation in seafood: A detailed review. Comprehensive Reviews in Food Science and Food Safety. 2014;**13**(6):1125-1140. DOI: 10.1111/1541-4337.12106

[60] Tongnuanchan P, Benjakul S. Essential oils: Extraction, bioactivities, and their uses for food preservation. Journal of Food Science. 2014;**79**(7): R1231-R1249. DOI: 10.1111/1750-3841.12492

[61] Prakash B, Kedia A, Mishra PK, Dubey NK. Plant essential oils as food preservatives to control moulds, mycotoxin contamination and oxidative deterioration of agri-food commodities e Potentials and challenges. Food Control. 2015;**47**:381-391. DOI: 10.1016/j.foodcont.2018. 01.018

[62] Rodriguez-Garcia I, Espinoza AS, Ortega Ramirez A, Leyva M, Siddiqui W, Cruz-Valenzuela R, et al. Oregano essential oil as an antimicrobial and antioxidant additive in food products. Critical Reviews in Food Science and Nutrition. 2016;**56**:1717-1727. DOI: 10.1080/10408398.2013.800832

[63] De Souza WFM, Mariano XM, Isnard JL, De Souza GS, De Souza Gomes AL, De Carvalho RJT, et al. Evaluation of the volatile composition, toxicological and antioxidant potentials of the essential oils and teas of commercial Chilean boldo samples. Food Research International. 2019;**124**:27-33. DOI: 10.1016/j.foodres.2018.12.059

[64] Sharma H, Mendiratta SK, Kant R, Gurunathan K, Kumar S, Pal T. Use of various essential oils as bio preservatives and their effect on the quality of vacuum packaged fresh chicken sausages under frozen conditions. LWT—Food Science and Technology. 2017;**81**:118-127. DOI: 10.1016/j.lwt.2017.03.048

[65] Dastgerdi GF, Goli SAH, Kadivar M. New antioxidant active film based on HDPE and peppermint essential oil for packaging soybean oil. Journal of the American Oil Chemists' Society. 2016;**93**:657-664. DOI: 10.1007/s11746-016-2806-9

[66] Mezza GN, Borgarello AV, Grosso NR, Fernandez H, Pramparo MC, Gayol MF. Antioxidant activity of rosemary essential oil fractions obtained by molecular distillation and their effect on oxidative stability of sunflower oil. Food Chemistry. 2018;**242**:9-15. DOI: 10.1016/j.foodchem.2017.09.042

[67] Okhli S, Mirzaei H, Hosseini SE. Antioxidant activity of citron peel (*Citrus medica* L.) essential oil and extract on stabilization of sunflower oil. Oilseeds & Fats Crops and Lipids. 2020;**27**:32. DOI: 10.1051/ocl/2020022

[68] Bensid A, Ucar Y, Bendeddouche B, Ozogul F. Effect of the icing with thyme, oregano and clove extracts on quality parameters of gutted and beheaded anchovy (*Engraulis encrasicholus*) during chilled storage. Food Chemistry. 2014;**145**:681-686. DOI: 10.1016/j.foodchem.2013.08.106

[69] Karoui R, Hassoun A. Efficiency of rosemary and basil essential oils on the shelf-life extension of atlantic mackerel (*Scomber Scombrus*) fillets stored at 2°C. Journal of AOAC International. 2017;**100**:335-344. DOI: 10.5740/jaoacint.16-0410

[70] Boskovic M, Glisic M, Djordjevic J, Starcevic M, Glamoclija N, Djordjevic V, et al. Antioxidative activity of Thyme (*Thymus vulgaris*) and Oregano (*Origanum vulgare*) essential oils and their effect on oxidative stability of

minced pork packaged under vacuum and modified atmosphere. Journal of Food Science. 2019;**84**(9):2467-2474. DOI: 10.1111/1750-3841.14788

[71] Ishkeh SR, Asghari M, Shirzad H, Alirezalu A, Ghasemi G. Lemon verbena (*Lippia citrodora*) essential oil effects on antioxidant capacity and phytochemical content of raspberry (*Rubus ulmifolius* subsp. sanctus). Scientia Horticulturae. 2019;**248**:297-304. DOI: 10.1016/j.scienta.2018.12.040

[72] Liu Z, Lin D, Shen R, Zhang R, Liu L, Yang X. Konjac glucomannan-based edible films loaded with thyme essential oil: Physical properties and antioxidant-antibacterial activities. Food Packaging and Shelf Life. 2021;**29**:100700. DOI: 10.1016/j.fpsl.2021.100700

[73] Hien LTM, Dao DTA. Black pepper essential oil nanoemulsions formulation using EPI and PIT methods. Journal of Food Processing and Preservation. 2021;**45**(3):e15216. DOI: 10.1111/jfpp.15216

[74] Wu M, Zhou Z, Yang J, Zhang M, Cai F, Lu P. ZnO nanoparticles stabilized oregano essential oil Pickering emulsion for functional cellulose nanofibrils packaging films with antimicrobial and antioxidant activity. International Journal of Biological Macromolecules. 2021;**190**:433-440. DOI: 10.1016/j.ijbiomac.2021.08.210

[75] Mahdi AA, Al-Maqtari QA, Mohammed JK, Al-Ansi W, Cui H, Lin L. Enhancement of antioxidant activity, antifungal activity, and oxidation stability of *Citrus reticulata* essential oil nanocapsules by clove and cinnamon essential oils. Food Bioscience. 2021;**43**:101226. DOI: 10.1016/j.fbio.2021.101226

[76] Ansarian E, Aminzare M, Azar HH, Mehrasbi MR, Bimakr M. Nanoemulsion-based basil seed gum edible film containing resveratrol and clove essential oil: In vitro antioxidant properties and its effect on oxidative stability and sensory characteristic of camel meat during refrigeration storage. Meat Science. 2022;**185**:108716. DOI: 10.1016/j.meatsci.2021.108716

[77] Kiralan SS, Karagoz SG, Ozkan G, Kiralan M, Ketenoglu O. Changes in volatile compounds of virgin olive oil flavored with essential oils during thermal and photo-oxidation. Food Analytical Methods. 2021;**14**:883-896. DOI: 10.1007/s12161-020-01926-w

[78] Calo JR, Crandall PG, O'Bryan CA, Ricke SC. Essential oils as antimicrobials in food systems: A Review. Food Control. 2014;**54**:111-119. DOI: 10.1016/j.foodcont.2014.12.040

[79] Dhifi W, Bellili S, Jazi S, Bahloul N, Mnif W. Essential oils' chemical characterization and investigation of some biological activities: A critical review. Medecines. 2016;**3**:1-16. DOI: 10.3390/medicines 3040025

[80] Falleh H, Jemaa MB, Saada M, Ksouri R. Essential oils: A promising eco-friendly food preservative. Food Chemistry. 2020;**330**:127268. DOI: 10.1016/j.foodchem.2020.127268

[81] Burt S. Essential oils: Their antibacterial properties and potential applications in foods: A review. International Journal of Food Microbiology. 2004;**94**:223-253. DOI: 10.1016/j.ijfoodmicro.2004.03.022

[82] Ben Jemaa M, Falleh H, Saada M, Oueslati M, Snoussi M, Ksouri R. *Thymus capitatus* essential oil ameliorates pasteurized milk quality. Journal of Food Science and Technology. 2018;**55**(9):3446-3452. DOI: 10.1007/s13197-018-3261-4

[83] Gutiérrez-del-Río I, Fernández J, Lombó F. Plant nutraceuticals as antimicrobial agents in food preservation: Terpenoids, polyphenols and thiols. International Journal of Antimicrobial Agents. 2018;**52**:309-315. DOI: 10.1016/j.ijantimicag.2018.04.024

[84] Adelakun OE, Oyelade OJ, Olanipekun BF. Use of essential oils in food preservation. In: Preedy VR, editor. Essential Oils in Food Preservation, Flavor, and Safety. London: Elsevier; 2016. pp. 71-84. DOI: 10.1016/C2012-0-06581-7

[85] Hussain MA, Sumon TA, Mazumder SK, Ali MM, Jang WJ, Abualreesh MH, et al. Essential oils and chitosan as alternatives to chemical preservatives for fish and fisheries products: A review. Food Control. 2021;**129**:108244. DOI: 10.1016/j.foodcont.2021.108244

[86] Han Lyn F, Nur Hanani ZA. Effect of Lemongrass (*Cymbopogon citratus*) essential oil on the properties of chitosan films for active packaging. Journal of Packaging Technology and Research. 2020;**4**(1):33-44. DOI: 10.1007/s41783-019-00081-w

[87] Amalraj A, Raj KKJ, HaponiukJT,ThomasS,GopiS.Preparation, characterization, and antimicrobial activity of chitosan/gum arabic/polyethylene glycol composite films incorporated with black pepper essential oil and ginger essential oil as potential packaging and wound dressing materials. Advanced Composites and Hybrid Materials. 2020;**3**:485-497. DOI: 10.1007/s42114-020-00178-w

[88] Fattahian A, Fazlara A, Maktabi S, Pourmahdi M, Bavarsad N. The effects of chitosan containing nano-capsulated *Cuminum cyminum* essential oil on the shelf-life of veal in modified atmosphere packaging. Journal of Food Measurement and Characterization. 2021. DOI: 10.1007/s11694-021-01213-0

[89] Langroodi AM, Nematollahi A, Sayadi M. Chitosan coating incorporated with grape seed extract and *Origanum vulgare* essential oil: An active packaging for turkey meat preservation. Journal of Food Measurement and Characterization. 2021;**15**:2790-2804. DOI: 10.1007/s11694-021-00867-0

[90] Sayadi M, Amiri S, Radi M. Active packaging nanocomposite gelatin-based films as a carrier of nano $TiO_2$ and cumin essential oil: The effect on quality parameters of fresh chicken. Journal of Food Measurement and Characterization. 2021. DOI: 10.1007/s11694-021-01169-1

[91] Sharma P, Ahuja A, Izrayeel AMD, Samyn P, Rastogi VK. Physicochemical and thermal characterization of poly (3-hydroxybutyrate-co-4-hydroxybutyrate) films incorporating thyme essential oil for active packaging of white bread. Food Control. 2022;**133**:108688. DOI: 10.1016/j.foodcont.2021.108688

[92] Caponio F, Durante V, Varva G, Silletti R, Previtali MA, Viggiani I, et al. Effect of infusion of spices into the oil vs. combined malaxation of olive paste and spices on quality of naturally flavoured virgin olive oils. Food Chemistry. 2016;**202**:221-228. DOI: 10.1016/j.foodchem.2016.02.005

[93] Yilmazer M, Karagöz SG, Ozkan G, Karacabey E. Aroma transition from rosemary leaves during aromatization of olive oil. Journal of Food and Drug Analysis. 2016;**24**:299-304. DOI: 10.1016/j.jfda.2015.11.002

[94] Sousa A, Casal S, Malheiro R, Lamas H, Bento A, Pereira JA. Aromatized olive oils: Influence of

flavouring in quality, composition, stability, antioxidants, and antiradical potential. LWT—Food Science and Technology. 2015;**60**(1):22-28. DOI: 10.1016/j.lwt.2014.08.026

[95] Perestrelo R, Silva C, Silva P, Câmara JS. Global volatile profile of virgin olive oils flavoured by aromatic/medicinal plants. Food Chemistry. 2017;**227**:111-121. DOI: 10.1016/j.foodchem.2017.01.090

[96] Baiano A, Gambacorta G, La Notte E. Aromatization of olive oil. Transworld Research Network. 2010;**661**:1-29

[97] Arcoleo G, Indovina MC, Varvara G, Lanza CM, Mazzaglia A. Improving olive oil shelf life with lemon essential oil. Chemical Engineering Transactions. 2009;**17**:849-854. DOI: 10.3303/CET0917142

[98] Wang D, Fan W, Guan Y, Huang H, Yi T, Ji J. Oxidative stability of sunflower oil flavored by essential oil from *Coriandrum sativum* L. during accelerated storage. LWT—Food Science and Technology. 2018;**98**:268-275. DOI: 10.1016/j.lwt.2018.08.055

[99] Moustakime Y, Hazzoumi Z, Joutei KA. Aromatization of virgin olive oil by seeds of *Pimpinella anisum* using three different methods: Physico-chemical change and thermal stability of flavored oils. Grain and Oil Science and Technology. 2021;**4**:108-124. DOI: 10.1016/j.gaost.2021.07.001

[100] Porto CD, Decorti D. Analysis of the volatile compounds of aerial parts and essential oil from *Thymus serpyllum* L. cultivated in North East Italy by HS-SPME/GC-MS and evaluation of its flavouring effect on Ricotta cheese. Journal of Essential Oil Bearing Plants. 2012;**15**:561-571. DOI: 10.1080/0972060X.2012.10644089

[101] Benkhoud H, M'Rabet Y, Ali MG, Mezni M, Hosni K. Essential oils as flavoring and preservative agents: Impact on volatile profile, sensory attributes, and the oxidative stability of flavored extra virgin olive oil. Journal of Food Processing and Preservation;**2021**:e15379 10.1111/jfpp.15379

Section 2

# Essential Oils and Food Science Technology

Chapter 4

# Essential Oils as Antimicrobial and Food Preservatives

*Mamdouh S. Serag, Reham A. Elfayoumy and Marwa T. Mohesien*

## Abstract

Essential oils (EOs) are secondary metabolites produced by aromatic and medicinal plants. These oils have a wide range of applications in the culinary, perfume, antimicrobial and food industries. Because of several reported side effects of synthetic oils, the use of essential oils as antimicrobials and food preservatives is a source of concern. For cereals, grains, pulses, fruits, and vegetables, essential oils have the potential to be employed as a food preservative. When compared to synthetic compounds, EOs derived from safe natural sources and are effective for human health. This chapter will shed light on some medicinal plants that are rich in essential oils, as well as their antimicrobial properties. Because essential oils are rich in a number of active ingredients [e.g., terpenes, terpenoids, carotenoids, coumarins, curcumins] that are important in food industry, they have strong antimicrobial and food preservation. As a result of the diverse properties of essential oils, they can be used in a natural, safe, eco-friendly, cost-effective and renewable manner. Examples of some foodborne diseases will also be highlighted.

**Keywords:** essential oils, antibacterial, antifungal, bioactivity, foodborne diseases, food preservatives

## 1. Introduction

Essential oils [EOs] are a volatile mixture of chemical molecules with a strong odour that is extracted from aromatic and medicinal plants. Steam or hydro-distillation or Soxhlet extraction [solvent extraction or continuous extraction] procedures are used to extract EOs from aromatic and medicinal plants [1]. Essential oil is a liquid that is extracted from flowers, leaves, bark, stems and roots by steam or water distillation. Essential oils are not at all oily-feeling, despite the word 'oil'being used. The majority of essential oils are clear, but some, like patchouli, orange and lemongrass, are amber or yellow. They are rich in chemicals like phenols, monoterpenes and ketones. These plant chemicals are called plant 'essences 'referring to the fact that they carry some of the plant's natural ability to resist bacteria and fungi. These are the same chemical molecules that have been isolated or synthesised by the pharmaceutical industry to make drugs.

 IntechOpen

Commercial antimicrobial treatments had been used to prevent food deterioration or contamination since ancient times. As a result of consumer concerns about synthetic preservatives, natural antimicrobials such as essential oils are receiving more attention. Essential oils and their components from aromatic and medicinal plants had been shown to have antibacterial, antifungal and food preservation properties against a variety of pathogenic microorganisms [2].

Essential oils are hydrophobic liquids of aromatic compounds that are volatile and oily in nature and present in various plant parts such as flowers, leaves bark, stem, root and seed. Many plant essential oils are useful as a flavour or aroma enhancer as food additives. Applications of essential oil that can act as antimicrobial agents are growing due to the broad range of activities, natural origins and are safe. Currently, essential oils are frequently studied for their antibacterial and antifungal [3] as well as for their use as food preservatives [4].

Essential oils are considered to be secondary metabolites and important for plant defence as they often possess antimicrobial properties [5]. The antibacterial properties of secondary metabolites were first evaluated using essential oil vapours [6]. Since then, essential oils or their components had been shown to not only possess broad-range antibacterial and antifungal properties [7, 8].

Wherever you buy essential oils, the quality might vary dramatically from one dealer to the next. Despite the fact that they are all essential oils, they all do not have the same medicinal potential. Furthermore, the price charged does not always reflect the quality of the vendor's oils. There are a few:

1. $CO_2$ extracts and absolutes are distilled in different manners.

2. Poor quality oils that have been distilled from poor crops, been handled improperly, are old, etc.

3. Adulterated oils that have chemicals or other oils added to them. These can cause harmful side effects, or at best, provide the only minimal therapeutic benefit of good quality oils.

The information of this chapter was extracted from the accessible international electronic databases [PubMed, Springer, Science Direct, Wiley and Google] and books by keywords were namely: antibacterial, antifungal, bioactivity, essential oils, foodborne disease, food preservative properties.

The major aim of this chapter highlights the use of essential oils and their antibacterial, antifungal and food preservative properties in controlling fungi associated with food commodities. Some food-borne diseases also will be discussed.

## 2. Essential oils of some medicinal plants

### 2.1 Lemon essential oil

Lemons are one of the most popular citrus fruits in the world, and they are often used in cooking since they are high in vitamins. It also gives food a pleasing flavour and scent. Lemon oil's stimulating, soothing, carminative, anti-infection, astringent, detoxifying, antiseptic, disinfecting, sleep-inducing and antifungal characteristics contribute to its health advantages. The antibacterial chemical limonene, which belongs to the terpenes [monoterepenes] group, is found in Lemon essential oil.

Lemon oil [citrus lemon] includes d-lemonene chemicals, which have been researched for their impact on immunological function, lymphatic, circulatory, and digestive systems. It has antibacterial properties and can help white blood cells, phagocytes, and lymphocytes combat infection [9].

### 2.2 Ginger essential oil

As a member of the Zingiberaceae family, *Zingiber officinale Rosc. [ginger]* is widely used as a spice or medicinal plant in folk and traditional medicine. Rhizomes, the therapeutic portion of ginger, are utilised in traditional medicine to cure a variety of diseases [10]. It is a volatile substance extracted by distillation of unpeeled rhizome of *Z. officinale* Roscoe plant. The essential oil of ginger has a strong, warm and spicy aroma. Its colour is clear to light amber, and its consistency becomes thicker with age and exposure to air. *Z. officinale* is well known for its medicinal and culinary properties. As turmeric and cardamom, ginger also belongs to the family of Zingiberaceae.

Neroli, clove, black pepper, rose [Rosa alba], turmeric, angelica, spikenard, cardamom, clary sage, sweet marjoram, fennel, jasmine, grapefruit, coriander seed, lemongrass Analgesic, antibacterial, anti-emetic, anti-inflammatory, antioxidant, antispasmodic, antitussive, aperitif, aphrodisiac, deputative, stimulant, laxative, febrifuge, digestive, expectorant, immune-modulatory, rubefacient, stomachic and sudorific qualities are all found in ginger oil [11].

### 2.3 Peppermint essential oil

When dispersed in the air, peppermint is one of the most effective essential oils for destroying respiratory tract pathogens. It works well as a topical application as well. Peppermint menthol is increasingly commonly found in sports creams and chest rubs, such as Halls Mentholyptus cough drops. Although the oil is effective at opening sinus passages, it should be used with caution in this regard.

Peppermint oil should be a part of every traveller's first aid kit. It can work wonders for motion sickness and general nausea for some people. An excellent digestive tonic, peppermint essential oil can soothe many stomach complaints. For the traveller, its effectiveness in calming motion sickness can be of great help.

In addition, Peppermint oil had been demonstrated to be useful in lowering the symptoms of irritable bowel syndrome, a painful disorder of the intestines, in at least eight controlled investigations. Peppermint is deliciously stimulating to the mind, brightening and sharpening mental attention in addition to supporting the digestive system.

With all of these additional advantages, peppermint is an effective antimicrobial [12].

### 2.4 Rosemary

Rosemary (*Rosmarinus officinal* is L.) is a valuable essential oil plant from the Lamiaceae family. According to the evidence found by anthropologists and archaeologists, rosemary, was used in medicine and food industry [13].

Rosemary is a popular spice and medicinal herb all over the world. Rosemary is usually regarded as one of the spices with the highest antioxidant capacity among natural antioxidants. Rosemary essential oils have antibacterial and antifungal properties. It is often used as a food preservative and condiment [14].

### 2.5 Thyme

Thymol is a natural volatile monoterpenoid phenol that is the main active ingredient of oil extracted from species *Thymus vulgaris* L., commonly known as thyme, and other plants such as *Ocimum gratissimum* L. and *Origanum spp.* L. Thyme Oil is one of the most antiseptic essential oils and is high in antioxidant rating. Thymol, a potent antibacterial, is the primary component of thyme oil. Aromatherapists are well aware that thyme essential oil is one of the most powerful antibacterial essential oils available. Thymol's antibacterial and antifungal properties had been well reported [15].

Thyme essential oils showed some of the strongest killing power against antibiotic-resistant bacteria, according to studies at the Western Infirmary, Glasgow, UK. Thyme oil kills the anthrax bacillus, the typhoid bacillus, meningococcus and the agent responsible for tuberculosis and is active against *Salmonella* and *Staphylococcus* bacteria.

### 2.6 Clove essential oil

*Syzygium aromaticum L.* is a member of the Myrtaceae family, which includes the myrtle, *Eucalyptus* and *guava* families, and has around 3000 species and 130–150 genera. Clove is a fragrant flower that is grown in Madagascar, Sri Lanka, Indonesia and China [16].

Clove may also be referred to as Clove Tree, Clove Bud, Clove Stem, Tropical Myrtle, Zanzibar Redhead, Cengkih, Chengkeh, Chingkeh. It is typically processed using steam- or hydro-distillation as a method for extracting oil from the flower buds, leaves and stems.

Clove oil comes from the flower buds and leaves of *S. aromaticum* also known as *Eugenia caryophyllata* tree. It has a strong spicy scent. It has an analgesic and stimulating effect. Clove stem and leaves essential oils are also available, however, due to its composition and aroma, essential oil produced from the buds is often preferred.

Clove bud essential oil contains up to 85% eugenol, a phenol that contributes significantly to the scent, medicinal effects and safety precautions. Clove buds essential oil also contains a variety of additional compounds, including the sesquiterpene B-caryophyllene, the esters Eugenyl acetate and B-caryophyllene [2].

### 2.7 Mustard essential oil

Mustard essential oil, which is frequently confused with mustard oil, is distilled from mustard seeds. Mustard essential oil is also known as mustard volatile oil. The essential oil includes 92 percent allyl isothiocyanate, the chemical that gives mustard its strong flavour. This allyl isothiocyanate, as well as key fatty acids including oleic acid, linoleic acid and erucic acid, contribute to mustard essential oil's lengthy list of medical properties [17].

Over the years, mustard oil had a mixed reputation in many regions of the world. It is utilised as an edible oil there and is believed to be highly healthful, although it is frequently considered poisonous, irritating and unfit for eating in the rest of the globe [18].

Mustard essential oil is one of the most potent essential oils available, and it can be used to cure a variety of diseases. This oil, which is extracted from the black seeds of mustard using the steam distillation method, had a variety of therapeutic characteristics that can help with a variety of health problems [19].

## 3. Anti-bacterial activity of essential oils

Pathogenic bacteria reduce the quality and quantity by 20–40% of the total harvest every year in grains, seeds, fruits and vegetables during cultivation, transportation and storage. *Clavibacter michiganensis*, *Pseudomonas syringae*pv, tomato, *Pseudomonas solanacearum, Pseudomonas cichorii*. Such bacteria cause substantial losses. There are many essential oils that had been evaluated for their potential for antibacterial activity against these pathogenic bacteria [20].

Gram-negative bacteria are generally more resistant to essential oils than Gram-positive bacteria. Gram-negative bacteria have hydrophilic lipopolysaccharides [LPS] in their outer membranes work as a barrier to macromolecules and hydrophobic chemicals, allowing them to tolerate hydrophobic antimicrobial substances such as those present in essential oils [21].

## 4. Anti-fungal activity of essential oils

Fungi can degrade food commodities such as grains, seeds, fruits and vegetables by producing mycotoxins, and they can make food unsafe for human consumption by lowering nutritional value [4]. Foodborne fungal infections and their toxic metabolites, according to the FAO, can cause qualitative and quantitative problems. Quantitative losses of up to 25% of total agricultural food commodities throughout the world [21].

Food quality, colour and texture are all reduced as a result of fungal infection in food commodities, as are the nutrients contained and the physiological aspects of food commodities. Fungi can create mycotoxins during infection, which can cause famines in underdeveloped nations [22].

Food contamination by *Alternaria, Aspergillus, Penicillium, Fusarium,* and *Rhizopus spp.* is an important issue in terms of moulds because of the associated health risks and foodborne diseases [4].

Essential oils have antifungal properties which are linked to the breakdown of fungal hyphae caused by mono- and sesquiterpene-molecules in the essential oils. Essential oils also increase membrane permeability, which means they can dissolve in cell membranes and produce swelling, limiting membrane function. Furthermore, essential oils' antifungal effect is due to their lipophilic feature, which allows them to enter cell walls and impact enzymes involved in cell-wall production, causing fungus to change their morphological traits [23].

## 5. Value of essential oils in food preservation

Essential oils have been successfully used in the preservation of food commodities in order to extend shelf life in recent years. Various researchers have employed essential oils, either in pure or formulation form, to extend the shelf-life of food commodities in a variety of storage containers, including those made of cardboard, tin, glass, polyethylene, or natural textiles, with positive results [4]. Essential oil constituents like citral, citronella, citronellol, eugenol, farnesol, and nerol among others, have been shown to protect chilli seeds and fruits from fungal infection for up to 6 months [24]. *Ageratum conyzoides* essential oil successfully stopped blue mould from destroying mandarins and extended their shelf life by up to 30 days [4].

Essential oils from *Cymbopogon nardus, C. flexuosus* and *Ocimum basilicum* were observed that could significantly control anthracnose in bananas and increased banana shelf-life by up to 21 days. For up to 3 weeks, *Cymbopogon flexuosus* essential oil [20 L/mL] can preserve *Malus pumilo* fruits from decaying. The use of *Cymbopogon pendulous* essential oil as a fumigant increased groundnut shelf-life by 6–12 months [25, 26], thus proving to be more effective than *Parkia roxburghii* essential oil. These differences in efficacy of essential oils may be related to the use of oils from different plant species, as well as to their chemical composition, dose level, and storage container type [25, 26].

Thyme (*Thymus capitate*) [0.1%] and Mexican lime (*Citrus aurantifolia*) [0.5%] oil reduced disease incidence in papaya fruit whereas, cinnamon [0.3%] oil increased banana storage life by up to 28 days and reduced fungal disease incidence in banana [27].

## 6. Foodborne diseases

The definition of food spoilage can be interpreted as the process in which food deteriorates to the point in which it is not edible to humans, this occurred by many spoilage microorganisms [bacteria and fungi], which by many reactions change the composition of food and deteriorates its texture, odour, colour and taste which make it unfit for human consumption [28].

Food diseases or food spoilage are widespread health problems and a major cause of the reduction in economic productivity and human lives around the world [28]. Food poisoning and spoilage are two different things, which affect the final quality and safety of foods. Food poisoning can be also referred to as food-borne illness. Many different forms of food-borne pathogens, such as bacteria and fungi, cause it when people eat contaminated food. Foodborne infections have become a major problem in the modern world since packaged food consumption has risen dramatically. Pathogens that penetrate packaged foods have a higher chance of surviving, which must be monitored. For this reason, antimicrobial chemicals are applied to food or packaging materials, either alone or in combination [28].

Foodborne infections are caused by pathogenic bacteria, fungus, and parasites infected [29]. Food safety is a well-known problem worldwide. This problem affects hundreds of millions of people who are injured by contaminated or spoiled food. 'One of the most widespread health concerns and a major cause of productivity loss and bad impact on human health, 'according to the World Health Organisation [28].

Intoxication, infection and toxic infection are three types of food contamination. Intoxication refers to the production of toxins after ingestion of harmful microorganisms in food; the microbe that produced and excreted the toxic waste products into the food may be killed, but the toxin they produced causes illness or digestive upset; toxic infection refers to the production of toxins after ingestion of harmful microorganisms in food; food infection is the other type of foodborne illness; It is caused by eating food that contains certain types of live microbe which are present in the food, once the food is consumed, the bacterial cells themselves continue to grow and illness can result. Symptoms of food poisoning are headaches, vomiting, nausea, diarrhoea and dehydration, and these symptoms can be out of control which can be fatal many times.

Consumers are increasingly concerned about the rising number of illnesses linked to harmful and spoilage microbes found in food. Food-borne infections affect millions of people every year all over the world, and they can vary from minor irritations to life-threatening conditions. Clinical microbiology laboratories play a critical role in the detection of these illnesses by identifying and reporting infections to public health officials, who then utilise the information to track down food-borne outbreaks [30]. According to the Center for Disease Control and Prevention [CDC], 76 million cases of food-borne disease occur in the United States each year [31]. The case figures are based on reportable disorders that each laboratory is obligated to report to their local or state public health officials, as well as active surveillance undertaken by the Center for Disease Control and Prevention (CDC). Pathogenic *E. coli*, *Campylobacter spp*. and *Salmonella* were the leading causes. Species., although the causes of approximately 80% of illnesses were unknown. Approximately 25% of the 15.9 million gastroenteritis episodes that occur in Australia are thought to be spread by contaminated food. This translates to an average of one foodborne gastroenteritis episode per five years per person [31].

### 6.1 Pathogens that cause foodborne disease

There were almost 250 distinct food-borne illnesses identified. The majority of these illnesses are infections caused by bacteria, fungus and parasites that can be spread by food. Poisonings, for example, are diseases caused by hazardous poisons or compounds contaminating food, such as toxic mushrooms or fungus. These different diseases have many different symptoms, so there is no one 'syndrome 'that is a food-borne illness. However, the microbe or toxin enters the body through the gastrointestinal tract, and often causes the first symptoms there, so nausea, vomiting, abdominal cramps and diarrhoea are common symptoms in many foodborne diseases, and if the symptoms are not controlled lead to be fatal [29].

#### *6.1.1 Listeria monocytogenes*

*L. monocytogenes* is a bacterium that causes food contamination which is responsible for listeriosis. It usually produces just a mild sickness in healthy people. *L. monocytogenes* can be found all over habitats. It had been isolated from domestic and wild animals, birds, soil, plants, feed, water and food processing factory floors, drains and damp places. This bacterium is a Gram-positive rod-shaped bacterium that do not generate spores. A lot of factors influence the growth and survival of *L. monocytogenes* [32, 37].

#### *6.1.2 Bacillus subtilis*

*B. subtilis* cells are rod-shaped, Gram-positive bacteria that are naturally found in soil and plants. *B. subtilis* grow in the mesophilic temperature range. The optimal temperature for growth ranging is 25–35°C. The creation of stress-resistant endospores is one such technique. Another strategy is the uptake of external DNA, which allows the bacteria to adapt by recombination. These solutions, however, take time to implement it can sometimes contaminate food, however, they seldom result in food poisoning, but it main responsible for many types of food spoilage. Several such species have been described which are mostly the variants of *B. subtilis*, they are probably present in most bread. *B. subtilis* had been reported spoiling canned seafoods, meats etc. [32].

#### *6.1.3 Micrococcus luteus*

*M. luteus* is a Gram-positive, coccoid bacterium [0.5 to 3.5 microns in diameter]. It is capable of dividing into more than one plane. *M. luteus* can be found on human skin as well as in soil, dust, water, and air. It is a typical part of the human body's flora. The bacterium colonises the human mouth, mucosae, oropharynx and upper respiratory tract. It is not usually considered a pathogen, or disease-causing organism of healthy people [33].

#### *6.1.4 Shiga toxin-producing Escherichia coli [STEC]*

*E. coli* are bacteria that form part of the normal gut flora of humans and other warm-blooded animals. Although most E. coli are considered harmless, certain strains can cause severe illness in humans, particularly Shiga toxin-producing *E. coli* [STEC], which is also known as verocytotoxin-producing *E. coli* [VTEC]. Infection with STEC is the main cause of haemolytic uraemic syndrome, a condition that can be fatal in humans. *E. coli* are Gram-negative, rod-shaped bacteria and are members of the family Enterobacteriaceae [34].

#### *6.1.5 Aspergillus flavus*

*Aspergillus* is a common mould found on bread and other types of food such as meat and fish, as the mould grows on food it produces enzymes that break down the food resulting in spoilage. In addition to enzymes, *Aspergillus flavus* also produce mycotoxins onto the food. Ingestion of mycotoxin-contaminated food is fatal. Hundreds of people in developing countries die every year after consuming grains contaminated with mycotoxins [35].

#### *6.1.6 Rhizopus stolonifera*

*Rhizopus sp.* is a genus of common saprophytic fungi on plants and specialised parasites on animals. The bread mould, *Ranunculus stolonifer*, may grow on a broad variety of foods and plants, causing food spoilage and plant diseases in the field. It thrives in somewhat acidic environments, thus it like both fruit and bread. Off-flavors, mycotoxins contamination, discoloration, and rotting are all symptoms of food spoilage caused by mould. Spoilage can occur either in the field or in storage. The water activity of the food determines the types of mould spoiling the food [36].

## 7. Value of essential oils in food preservation

In recent years, there has been successful research into the use of essential oils in the preservation of food commodities in order to extend shelf life. Various investigators had used essential oils, either in pure or formulation forms, to enhance the shelf-life of food commodities in different storage containers such as those made of cardboard, tin, glass, polyethylene, or natural fabrics and have observed significant enhancement of shelf-life [4]. An earlier study reported that some essential oil constituents such as citral, citronella, citronellol, eugenol, farnesol and nerol could protect chilli seeds and fruits from fungal infection for up to 6 months [24]. Essential oil from *A. conyzoides* successfully controlled the rotting of mandarins by blue mould and increased mandarin shelf-life by up to 30 days [37]. Essential oils from *C. nardus, C. flexuosus* and *O. basilicum* and observed that they could significantly control anthracnose in banana and increased banana shelf-life by up to 21 days. For up to 3 weeks, *C. flexuosus* essential oil [20 L/mL] can prevent *Malus pumilo* fruits from decaying [26, 38]. A fumigant application of essential oils from *Putranjiva roxburghii* was effective against *A. flavus* and *A. niger* infecting groundnuts during storage and enhanced the shelf-life of groundnut from fungal biodeterioration for up to 6 months [24]. The use of *Cymbopogon pendulous* essential oil as a fumigant increased groundnut shelf-life by 6–12 months, thus proving to be more effective than *P. roxburghii*. Food preservation involves preventing the growth of foodborne microorganisms that lead to food spoilage or food contamination just like bacteria, fungi [39]. Unpreserved food can lead to its spoilage by microorganisms which make denaturation to the food and make it unfit to be consumed so the economy of the countries will be affected, or the food can be contaminated by dangerous pathogenic foodborne microorganisms [25, 40].

## 8. Conclusion

Researchers from all over the world had been drawn to the study of plant antimicrobials as a result of their work on essential oils. Essential oils and their compounds have clearly been extensively characterised, and essential oils had been employed to combat a wide range of diseases. As a result, this chapter included a brief summary of essential oils and how they can be used as antibacterial, antifungal, and food preservatives. Essential oils have a wide spectrum of antibacterial characteristics, according to the relevant literature review and their natural sustainability when they are utilised as possible biocontrol agents against pathogenic bacteria and fungi. In this regard, we suggest that essential oils are safe and cheap as biocontrol products that should be investigated further because of their ability to preserve food.

## Author details

Mamdouh S. Serag*, Reham A. Elfayoumy and Marwa T. Mohesien
Faculty of Science, Department of Botany and Microbiology, Damietta University, Egypt

*Address all correspondence to: mamdouhserag054@gmail.com; mserag@du.edu.eg

## References

[1] Raut JS, Karuppayil SM. A status review on the medicinal properties of essential oils. Industrial Crops and Products. 2014;**62**:250-264

[2] Haro-González JN, Castillo-Herrera GA, Martínez-Velázquez M, Espinosa-Andrews H. Clove essential oil (*Syzygium aromaticum* L. Myrtaceae): Extraction, chemical composition, food applications, and essential bioactivity for human health. Molecules. 2021;**26**(6387):1-25. DOI: 10.3390/molecules26216387

[3] Nedorostova L, Kloucek P, Kokoska L, Stolcova M, Pulkrabek J. Antimicrobial properties of selected essential oils in vapour phase against foodborne bacteria. Food Control. 2009;**20**:157-160

[4] Pandey AK, Kumar P, Singh P, Tripathi NN, Bajpai VK. Essential oils: Sources of antimicrobials and food preservatives. Frontiers in Microbiology. 2017, 2017;7:2161. DOI: 10.3389/fmicb.2016.02161

[5] Tajkarimi MM, Ibrahim SA, Cliver DO. Antimicrobial herb and spice compounds in food. Food Control. 2010;**21**:1199-1218

[6] Burt S. Essential oils: Their antibacterial properties and potential applications in foods a review. International Journal of Food Microbiology. 2004;**94**:223-253

[7] Deans SG, Ritchie G. Antibacterial properties of plant essential oils. International Journal of Food Microbiology. 1987;**5**:165-180

[8] Silva F, Ferreira S, Duarte A, Mendonça DI, Domingues FC. Antifungal activity of *Coriandrum sativum* essential oil, its mode of action against Candida species and potential synergism with amphotericin B. Phytomedicine. 2011;**19**:42-4710

[9] Available from: https://www.organicfacts.net/health-benefits/essential-oils/health-benefits-of-lemon-oil.html

[10] Mahboubi M. Zingiber officinale Rosc. Essential oil, a review on its composition and bioactivity. Clinical Phytoscience. 2019;**5**(6):1-12. DOI: 10.1186/s40816-018-0097-4

[11] Available from: https://www.everphi.com/2018/08/19/ginger-essential-oil/

[12] Available from: https://mightynest.com/articles/the-essential-6-uses-for-peppermint-essential-oil

[13] Erkan IE, Aşcı ZA. Studies on antimicrobial, antifungal and antioxidant properties of rosemary. Turkish Journal of Agriculture - Food Science and Technology. 2020;**8**(12):2708-2715

[14] De Oliveira J.R. , Camargo, S.E. and Dias de Oliveira, L. 2019. *Rosmarinus officinalis* L. (rosemary) as therapeutic and prophylactic agent. Journal of Biomedical Science, 26 [5]: 1-22. 10.1186/s12929-019-0499-8

[15] Escobar A, Perez M, Romanelli G, Blustein G. Thymol bioactivity: A review focusing on practical applications. Arabian Journal of Chemistry. 2020;**13**:9243-9269

[16] Golmakani MT, Zare M, Razzaghi S. Eugenol enrichment of clove bud essential oil using different microwave-assisted distillation methods. Food Science and Technology Research. 2017;**23**:385-394

[17] Available from: https://www.organicfacts.net/health-benefits/essential-oils/health-benefits-of-mustard-essential-oil.html

[18] Available from: https://www.healthmagazine.ae/health-tips-remedies/health-benefits-mustard-essential-oil/

[19] Available from: https://www.environment.co.za/alternative-healing-natural-remedies/oil-of-mustard-one-of-the-strongest-essential-oil-available.html

[20] Agrios G. Plant Pathology. 5th ed. Amsterdam: Elsevier Academic Press; 2005

[21] Trombetta D, Castelli F, Sarpietro MG, Venuti V, Cristani M, Daniele C, et al. Mechanisms of antibacterial action of three monoterpenes. Antimicrob.Agents Chemother. Antimicrobial Agents and Chemotherapy. 2005;**49**(6):2474-2478. DOI: 10.1128/AAC.49.6.2474-2478.2005

[22] Wagacha JM, Muthomi JW. Mycotoxin problem in Africa: Current status, implications to food safety and health and possible management strategies. International Journal of Food Microbiology. 2008;**124**:1-12. DOI: 10.1016/j.ijfoodmicro.2008.01.008

[23] Cox P, Betts R, Jones C, et al. Acceleration of global warming due to carbon-cycle feedbacks in a coupled climate model. Nature. 2000;**408**:184-187. DOI: 10.1038/35041539

[24] SKM T. Palynostratigraphy Zonation and correlation of the Jowai-Sonapur Road section, [Palaeocene—Eocene], Meghalaya, India. In: Tiwari RS, editor. 5th Proceddings of Indian Geophytological Conference, Special Publication. New Delhi, India: ISA; 1984. pp. 316-328

[25] Shukla A. Stimulation of mRNA export by an F-box protein, Mdm30p, *In vivo*. Journal of Molecular Biology. 2009;**389**(2):238-247

[26] Antony M, Shukla Y, Janardhanan KK. Potential risk of acute hepatotoxicity of kodo poisoning due to exposure to cyclopiazonic acid. Journal of Ethnopharmacology. 2003;**87**(2-3):211-214

[27] Maqbool SB, Mehrotra S, Kolpakas A, Durden C, Zhang B, Zhong H, et al. Dampened activity of E2F1-DP and Myb-MuvB transcription factors in Drosophila endocycling cells. Journal of Cell Science. 2010;**123**(23):4095-4106

[28] Available from: https://www.who.int/health-topics/foodborne-diseases

[29] Center for Disease Control and Prevention (CDC). Foodborne Illness - Frequently Asked Questions. Available from: http://www.cdc.gov/foodborneburden. Clifton Road Atlanta, USA: US Centers for Disease Control and Prevention; 2011

[30] Linscott AJ. Food-borne illnesses. Clinical Microbiology Newsletter. 2011;**33**(6):41-45

[31] Mead PS, Slutsker L, Dietz V. 1999, Food-related illness and death in the United States. Emerging Infectious Diseases;(5, 5):607-625. DOI: 10.3201/eid0505.990502

[32] Available from: http://www.gitam.edu/eresource/environmental/em_maruthi/food.htm

[33] Available from: https://www.britannica.com/science/Micrococcus

[34] Meng J, Schroeder CM. Escherichia coli. Chapter 1. In: Simjee S, editor.

Foodborne Diseases. Totowa: Humana Press; 2007. pp. 1-25

[35] Nielsen PV, Rios R. Inhibition of fungal growth on bread by volatile components from spices and herbs, and the possible application in active packaging, with special emphasis on mustard essential oil. International Journal of Food Microbiology. 2000;**60**:219-229

[36] Kungu J. Mould and food spoilage. Journal of Microbiology. 2005;**9**(4):177-188

[37] Lado B, Yousef AE. In: Ryser ET, Marth EH, editors. Characteristics of *Listeria Monocytogenes* Important to Food Processors. Chapter 6. Boca Raton: CRC Press Taylor & Francis Group; 2007. pp. 157-213

[38] Dixit V, Pandey V, Shymar R. Differential antioxidativeresponses to cadmium in roots and leaves of pea [*Pisum sativum*L. cv Azard]. Journal of Experimental Botany. 2001;**52**:1101-1109

[39] Anthony JL, Lonigan CJ, Driscoll K, Phillips BM, Burgess SR. Phonological sensitivity: A quasi-parallel progression of word structure units and cognitive operations. Reading Research Quarterly. 2003;**38**:470-487

[40] Available from: https://en.wikipedia.org/wiki/Food_preservation

Chapter 5

# Encapsulation of Essential Oils and Their Use in Food Applications

*Hamdy A. Shaaban and Amr Farouk*


## Abstract

Due to the modern lifestyle and consumers' interests, demands toward healthy foods and nutraceuticals were increased, among them essential oils (EOs) characterized by different biological activities. However, the use of EOs in foods and pharmaceuticals may be limited due to the hydrophobicity nature in addition to the instability and cause of degradation upon exposure to environmental conditions, e.g., oxygen, temperature, and light. Therefore, encapsulation in various colloidal systems such as microcapsules, nanospheres, nanoemulsions, liposomes, and molecular inclusion complexes, seem to be the solution for such issues. New trends in food packaging have also been focused on exploiting capsulated bioactive EOs constituents for extending foods' shelf life due to their potent antimicrobial agents and the great activity against pathological bacteria. Micro and nanoencapsulation of EOs may affect their biological activities based on the technique used. In the current chapter, different subjects have been discussed, like techniques used for the encapsulation of EOs, potential applications in food, and their behaviors/trends after encapsulation.

Moreover, the benefits of encapsulation, namely bioavailability, controlled release, and protection of EOs against environmental stresses, are discussed. The applications of encapsulated EOs are also summarized in this chapter. Also, the relevance of the encapsulation of EOs as antimicrobial agents and their incorporation into food packaging are discussed.


**Keywords:** essential oils, encapsulation, biological activities, food preservation

## 1. Introduction

Essential oils (EOs) can be extracted from any part of plants and are considered secondary metabolites. They usually comprise a complex mixture of alkaloids, flavonoids, isoflavones, monoterpenes, phenolic acids, carotenoids, and aldehydes [1]. EOs consist of a broad spectrum of components in which the efficacy as antimicrobial, antioxidants, etc., comes from the synergistic effect of many components. These components are responsible for the ability of EOs to be introduced and incorporated in many applications, such as in cosmetics, nutraceuticals, and food products. The application of EOs industrially is often limited. They are susceptible to environmental conditions such as light, oxygen, and temperature; they easily evaporate, are nearly insoluble in water, and have strong lipophilicity and volatility [2]. As a result, exploring the potential to extend their applications has become a key research issue.

Encapsulation has been introduced to improve EOs applications. It allows for the preservation of bio-functional properties of EOs, enhancing their stability against harsh conditions, giving benevolent masking effect, and providing controlled release of EOs. In a study by Shetta et al. [3], it was found that encapsulation significantly enhances the thermal stability of encapsulated peppermint and green tea EOs around 2.18 and 1.74 folds, respectively, pure EOs. Encapsulation can be achieved by many techniques and divided into 1) chemical method, 2) physicomechanical method, and 3) physicochemical method. The encapsulation process might involve more than one technique [4]. The selection of the most feasible technique would depend on the type of coated material, the operational cost, and the application of the encapsulation products. Encapsulation parameters such as encapsulation efficiency, encapsulation yield, payload/loading capacity, and surface loading are commonly used as primary indicators to reflect the performance of the encapsulation process and quality of encapsulation products (encapsulates).

Packaging protects foods from environmental factors and microbial contamination to maintain food quality and safety [5]. Using bioactive packaging avoids food spoilage and poisoning, which seriously affects public health and extends the shelf life of food products, especially those susceptible to microbial spoilage [6]. Unlike routine packaging, which only avoids the exchanges of air gases, moisture, and aromatic compounds between the food and the environment around [7], bioactive packaging provides antimicrobial activity to extend shelf life and food safety [8].

The safety and quality of packaged food by incorporating natural antimicrobial compounds and natural antioxidant compounds [9] is now an active research area [10–14]. Unfortunately, their use in raw form in food packaging materials is restricted by the hydrophobicity nature and the low stability against the environmental conditions during the processing, distribution, and storage of foods [15]. Also, the uncontrolled release of volatile active constituents of EOs can significantly negatively affect their biological benefits [15]. To overcome such limitations, appropriate carriers and encapsulation techniques were designed.

The design of the encapsulation method on the form of essential-oil-loaded particles is a complex process with interrelated steps [5] based on many factors like choosing the wall material, technique used, and the intended matrix in which essential oils are to be incorporated [8]. Basically, the nanoencapsulation process is the coating or trapping EOs as a core material by biopolymers to avoid the limitations of using EOs as a natural food preservative. Accordingly, different techniques could be used for the nanoencapsulation of EOs, such as nanoemulsion and liposomes. In the specific case of essential oil nanoemulsion, the preparation consists of a biphasic liquid system of one liquid solution dispersed in a continuous medium, and no polymer shells are used [12]. The presence of EOs in stable nanoemulsions helps enhance their dispersibility in aqueous solutions, avoid the interaction with other food ingredients or environment, keep their organoleptic properties, and improve their absorption and bioavailability. Therefore, nanoemulsions ofEos as a natural powerful food conservator became a potential target with respect to the encapsulation technique, leading to the instability or the inefficiency of the produced emulsion. A better understanding of the EOs encapsulation phenomenon would widen the knowledge of possible alternatives to consider while designing green food preservatives for future research. Accordingly, this chapter covers a general description of the EOs and encapsulation techniques along with evaluation for these methods and a comparison between nano- and microencapsulation. Finally, the effect of the nanoemulsion technique used on the EOs constituents was discussed based on recently published studies [16].

## 2. Essential oils

EOs are natural substances consisting of mixtures of different volatile and aromatic constituents. They are widely found in herbs and spices such as garlic, black cumin, cloves, cinnamon, thyme, basil, bay leaves, coriander, mustard, rosemary, sage, and others [17, 18]. The EOs constituents produced as secondary metabolites have many functions, such as insecticidal, antimicrobial agents, or attracting insects to help in flower pollination [18]. Flowers, leaves, stems, roots, fruits, and even seeds could be sources for EOs. Different techniques are used to extract, such as steam and hydro-distillation. Organic solvents extraction such as ethanol, acetone, and methanol are also used based on the polar solubility of the different constituents of EOs [19–21]. Distillation of EOs depends on their density, which is mostly less than 1, despite a few exceptions, e.g., cinnamon, sassafras, clove, and vetiver [22]. Based on the structure of their different constituents, the bioactivity of EOs showed various potential uses and applications as antimicrobial, antioxidant, and antifungal agents against yeasts and filamentous fungi, which represented a potential natural and potential healthy use as food preservatives [23–25]. Due to their antioxidant activity, the application of different EOs in food industries, especially fats and oils, to avoid lipid peroxidation caused by free radicals represented a potential target [26]. Lipid peroxidation results in many negative impacts for food products, including unpleasant aroma and flavors, deterioration of the food quality, decreasing the nutritional value of food, and severe health issues [27]. Based on the modern lifestyle and consumer demands, using EOs as natural antioxidants is favored over synthetic antioxidants such as butylated hydroxyanisole (BHA), butylated hydroxytoluene (BHT), and tert-butylhydroquinone whose applicability has been discouraged due to safety, health, and environmental concerns [28].

Among others, phenols, esters, terpenes, sesquiterpenes, aldehydes, ethers, alcohols, and phenylpropanoids, constitute the main classes in EOs responsible for bioactivity and sensory properties [22, 25]. For example, thymol, carvacrol, α-terpinene, eugenol have antioxidant effects [29], while other constituents such as limonene, eugenol, pinene, carvone, and linalool carvacrol have been suggested as agents responsible for the antimicrobial efficiency against foodborne pathogens [30]. In the same context, eugenol exhibited an efficient bactericidal activity against *Salmonella enterica* serovar *Typhimurium* as well as carveol, citronellol, and geraniol which presented anti bactericidal activity against E. coli, while terpineol had good activity versus S. aureus strains [31]. The presence of hydroxyl groups is responsible for the previously described compounds' higher antimicrobial activity [31]. Meanwhile, other compounds belonging to different classes such as benzoic acids, benzaldehydes, and cinnamic acid have shown up to 50% inhibition of *Listeria monocytogenes* under anaerobic conditions [32]. The EOs of similar plants have been reported to have differences in composition depending on the geographical location that the plant is found [33]. Notably, the composition and yield of EOs can vary with environmental conditions, climate, harvesting stages, planting, preparation methods, and genetics [34]. For example, weather parameters like rains and temperature have influenced the oil content and its constituents [35].

## 3. Encapsulation

Encapsulation of active ingredients or a core in solid walls or carriers represents a potential solution to control their release during storage or application and protect them from environmental conditions or interactions with the matrix around. As EOs are hydrophobic, emulsifying or dispersing in an aqueous solution represents the important primary step in the whole process. Following the emulsifying process, encapsulation can be performed by different techniques; chemical techniques like molecular inclusion or interfacial polymerization; physicochemical techniques like conservation and liposome encapsulation; and physical techniques like spray drying, spray chilling/cooling, co crystallization, extrusion, or fluidized bed coating [36]. Based on the technique and energy used, capsules can be found in micro and nanoscales, where microcapsules range between a few micrometers and a few millimeters while nanocapsules are found in the range of 53.8–415.2 nM. Other factors affect the size and physical properties of the capsules, like the natural or synthetic polymers used as wall materials and the core used. This system can increase the passive cellular absorption mechanisms, reduce mass transfer resistances, and increase antimicrobial activity due to its subcellular size [37].

## 4. Encapsulation of EOs

Natural EOs and extracts have limited applicability [38] because of drawback reactions during processing, transporting, or storage like oxidation, hydrolysis, crystallization, or enzymatic deterioration in the presence of oxygen and light [39, 40]. The lower thermal stability during food processing causes loss of EOs active components' biological functionalities [41] and significantly deteriorate their flavor and solubility. For example, the pomegranate peel extract associated with easy oxidation causes color deterioration and other instability issues [42], while Satureja hortensis EO drastically changes composition upon heating over 160°C [43]. The intense flavor of EOs, which is used as preservatives, may be transferred to the packed foods and negatively affect the final product's sensory properties [44, 45], so encapsulation is required to avoid the volatility of EOs bioactive components [46]. Consequently, many researchers have encapsulated them into other protection materials in order to make full use of their antioxidant and antimicrobial properties [47]. Nanoencapsulation of bioactive components to apply in food packaging materials are a potential target, growing steadily [48], since it can protect the components and therefore their biological efficiency against oxidative degradation upon exposure to air or high temperatures and during food processing [49, 50], in addition, to control their releasing [51]. For example, encapsulating thyme EO into cyclodextrin/ε-polylysine can reduce undesirable deficiencies such as volatility and hydrophobicity of its bioactive components [52]. While carvacrol, characterized by its antimicrobial activity, can be protected/encapsulated in a starch fiber matrix to avoid direct contact with food and reduce the effects on sensory features [53]. Encapsulation in zein microparticles improved the thermal stability of polyphenols from maqui fruit extract when exposed to high temperatures related to processed foods [54]. Orange and thyme oil adsorbed in halloysite or montmorillonite clay and then encapsulated in a polyethylene/polyamide/polyethylene multilayer film prolongated aroma release [55]. Encapsulation of black pepper (*Piper nigrum* L.) EO into sodium alginate and gelatin by complex coacervation avoids the loss of the main volatile from EOs preserved (80% of their original content) [56].

| Encapsulation method | Description | Nanoencapsulation | Microencapsulation |
|---|---|---|---|
| Emulsification | Emulsification is a process of mixing two immiscible solvents, and the resulting product is referred to as an emulsion. It can be divided into top-down approaches (high-shear stirring, high-pressure homogenization, microfluidization, and ultrasonication) and bottom-up (phase inversion temperature, emulsion phase inversion, and spontaneous nano emulsification) approaches. | Vitamin E encapsulated by Tween-80; vanillin encapsulated in poly(lactic acid) nanoparticles | Curcumin encapsulated by Tween 80 and polyglycerol polyricinoleate; lycopene encapsulated in plant (soy and pea) or dairy (whey and sodium caseinate) proteins |
| Spray drying | The basic theory of spray-drying is to feed the liquid into a drying chamber in the form of tiny droplets containing biologically active compounds, supplying hot air to the drying chamber, forming microcapsules in the drying chamber, and recovering them through a cyclone. | Folic acid encapsulated by whey proteins and resistant starch; curcumin encapsulated by chitosan/Tween 20 | Propolis extracts bioactive compounds encapsulated by maltodextrin matrices with or without nature gums; cocoa volatile compounds encapsulated by maltodextrins and modified starch |
| Freeze drying | The basic principle of freeze-drying is to freeze water contained in a solution or suspension and then evaporate the water molecules from the solution or suspension. | Fish oil encapsulated by poly-e-caprolactone and Pluronic F68 | Blackberry by-product extract encapsulated by maltodextrins; flaxseed oil encapsulated by sodium alginate, whey protein, and maltodextrin |
| Extrusion | Extrusion technique involves the injection of a bio-based solution into another solution to promote gelation and produce a hard and dense encapsulation system. | Seed oils encapsulated by sodium alginate and high methoxyl pectin | Canola oil encapsulated by alginate and high methoxyl pectin; quercetin encapsulated by carnauba wax, shellac, or zein |
| Complex coacervation | Coacervation is a well-known implemented technique to produce micro- and nanosystems. The basic mechanism is the formation of an emulsion by electrostatic attraction between oppositely charged molecules to produce the encapsulating structure. | Folic acid encapsulated by casein nanoparticles; anthocyanins encapsulated by whey protein isolate and beet pectin | Algal oil encapsulated by soy protein isolate and chitosan; β-carotene encapsulated by casein and gum tragacanth |
| Electro-spinning and electro-spraying | They are two modes of electrohydrodynamic processes that use a charged jet to rotate or spray a polymer solution to produce fibers or particles. | Rosehip seed oil encapsulated by zein prolamine fiber; β-carotene encapsulated by zein prolamine fiber | D-limonene encapsulated by seed gum and tween 20; fish oil encapsulated by a composite zein fiber |

**Table 1.**
*Summary of recent studies on micro- and nanoencapsulation of food bioactive compounds.*

## 5. Encapsulation process evaluation method

Generally in encapsulation, the idea of quantifying EO upon encapsulation process is 1) to calculate encapsulation efficiency and other encapsulation parameters; 2) to perform a controlled release study and understand the kinetics of release [57]; as well as 3) to evaluate the stability of encapsulates based on how much oil is left in the encapsulates [58], or how much oil is released to the releasing media [59], and still adhered to the surface [60]. Besides that, it is crucial to determine the components that are successfully encapsulated and responsible for the bio-function of EOs exactly. These components or types of EOs would have effects on encapsulation evaluation parameters. In a study by ref. [61], different encapsulation efficiency values were obtained when encapsulating kaffir lime oil from peels (KLO-P) and twigs oil fraction (KLO-TF) using chitosan as wall material. It was found that the encapsulation efficiency of KLO-TF is greater than KLO-P. The encapsulation efficiency difference was attributed to the components presented in each kaffir lime oil in which KLO-TF contains more oxygenated monoterpene components while the hydrocarbon monoterpenes components dominate KLO-P. Oxygenated monoterpenes components are more likely to interact with the functional group (active site) in the encapsulate, and as a result, more KLO-TF was successfully encapsulated.

Determination of EO in encapsulates can be done gravimetrically through direct measuring [62] or the distillation process. However, drawbacks associated with such techniques are that a large amount of formulation is required, improper extraction, and chances of loss of EO due to volatilization. To overcome these issues, reliable techniques using analytical methods such as chromatographic or spectrophotometric methods are introduced and expected to exhibit higher values than when the thermogravimetric analysis is used [63]. When employing these analytical methods, sometimes, digestion of the wall material is required to be achieved physically, chemically, or enzymatically [64]. **Table 1** below shows different types of EOs and commonly used solvents and methods to digest encapsulated walls. Subsequently, EO is extracted using an organic solvent such as hexane [65], petroleum ether, ethanol [66], or non-ionic surfactant; tween-80 before quantification using appropriate analytical methods. These analytical methods also have some disadvantages, such as possible experimental error, chances of loss of EO due to volatilization, and the possibility that the method selected is not convenient. For example, in cases where digestion of encapsulated walls is needed, the digested wall materials might somehow interfere with the spectrometric reading of EO. However, this could be resolved by using appropriate solvent and technique. Tolun et al. [66] used hexane to extract Moxa oil from encapsulates since gelatine and Gum Arabic used as encapsulating material did not interfere with the measurement process as they were insoluble in hexane. Meanwhile, Fraj et al. [67] used derivative spectrophotometry for quantitative analysis of core material since wall materials used (vitamin C and genipin) were also soluble in ethanol.

## 6. Nanoencapsulation versus microencapsulation

A comparison of micro- and nanoencapsulation functionality has been reported by [68], as shown in **Figure 1**. The main functionalities of microencapsulation taken into consideration are protecting active ingredients, including the extension of shelf

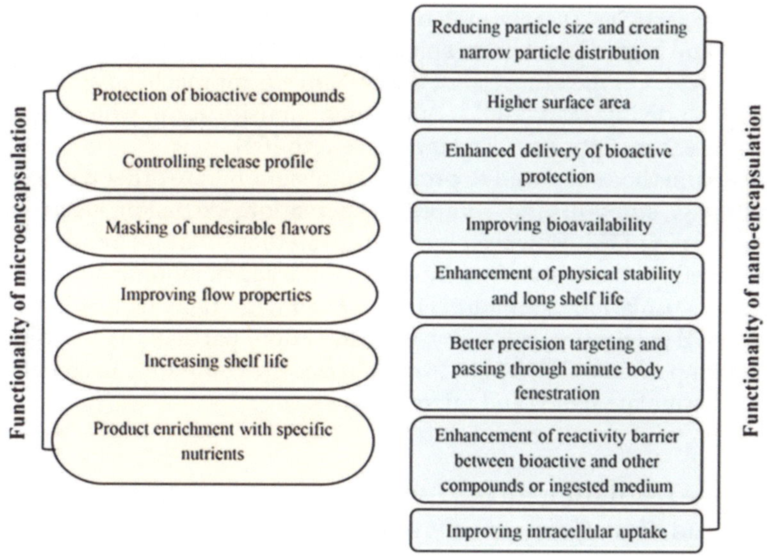


**Figure 1.**
*Advantagesofnano-andmicroencapsulation [69].*

life and controlling the release of bioactive components. While for nanocapsules, more attention is given to the functionals related to the size reduced like higher surface area and improving intracellular uptake. According to the authors, the formulation in nanoscales may improve bioavailability; however, this may depend on the technique used, as discussed later in section 7 of this chapter.

Particle size is an important factor affecting the functional characteristics of capsules [70]. Nanoencapsulation is the formulation of capsules with less than 1 micron (1000 nm), possessing different properties than ordinary encapsulation. According to the literature, capsules should be less than 100 nm to be considered nanocapsules [71]. The nanometric size of delivery systems can increase the surface area and, consequently, the food matrices' dispersion to form uniform and stable colloidal suspensions and may have better sustained-release effects than microcapsules. Based on their smaller size, nanocapsules can increase the passive cellular absorption mechanisms, promoting the effective release of active substances inside the target cells and consequently increasing the efficiency of active substances and their bioavailability.

Meanwhile, nanoparticles may penetrate the tissues (such as the liver) through the capillaries and are absorbed by the cells in the tissues; thus, the active substance can be efficiently delivered to the target cells in the body [72]. In the case of emulsions-based delivery systems, some interesting physical properties can distinguish nano and microemulsions. Microemulsions generally exhibit multiple scattering of visible light, which means they have an opaque white appearance. Conversely, nanoemulsions are much smaller than visible wavelengths, and therefore, they appear almost optically transparent, making them easily applied in the beverage industry [73].

Despite the numerous technologies for encapsulating biologically active compounds studied, only a few techniques, namely spray-drying and freeze-drying, are widely applied in the food industry [74]. Emulsification represents the first step of encapsulation. There are two types of approaches used to produce emulsions: a top-down approach and a bottom-up approach. The top-down approach involves reducing coarse particles' size through intensive mechanical destructive forces like high-pressure and high-shear homogenization, microfluidization, and microchannel homogenizers [75]. On the other hand, the bottom-up approach generally includes self-assembly, phase inversion, and spontaneous emulsification, which are affected by pH, temperature, concentration, and ionic strength [69].

Low-energy methods are used to prepare emulsions before other nanoencapsulation methods, e.g., spray-drying, complex coacervation, extrusion, electro-spinning, and electro-spraying [76]. However, low-energy methods require more stabilizers and surfactants to reduce the size and easily disperse the active ingredients [69]. Choosing the primary encapsulation technique is interrelated with many factors like the core and wall material properties, solubility, emulsification, particle size distribution, and food matrix composition [76]. **Table 1** summarizes the commonly used encapsulation techniques to formulate nano- and microcapsules.

## 7. Effect of Encapsulation by the Intensive-energy techniques on the structure and bioactivity of EOs components

Literature dealing with the encapsulation of EOs focused on the physical stability and biological activity of the micro or nanoparticles but not on the changes in the volatile constituents of the capsules. Few studies have reported that the formulation based on energy-intensive techniques like high-pressure and high-shear homogenization may lead to Ostwald ripening, flocculation, or coalescence of the emulsion with changes in its physical stability and biological activity [77]. Ali et al. [78] studied the effect of nanoencapsulation on volatile components and the bioactivity of Algerian *Origanum glandulosum* Desf. essential oil, a significant quantitative difference was observed in the level of monoterpenes between hydrodistilled oil and its nanocapsules. Additionally, the majority of sesquiterpenes were not detected in the nanocapsules extract. They owed that to the intensive-energy homogenization at 18000 rpm. Also, they reported that essential oil exhibited a higher antioxidant activity than nanocapsules and nanoemulsions, while nanocapsules showed the most potent cytotoxic effect on liver cancer cell line Hep-G2 in comparison to HD oil and nanoemulsions. In the same context, thymol and carvacrol were detected as predominates in the nanoemulsion of Algerian

*Saccocalyx satureioides* Coss. et Durieu oil was prepared by high-pressure homogenization, while borneol and α-terpineol were the major compounds detected in the same hydrodistilled oil, which affected the bioactivity of the oil and nanoemulsion [79]. Also, *Citrus sinensis* L.peel essential oil exhibited antifungal activity against *A. niger*, *A. ochraceus*, *Fusarium* spp., and *Penicillium* spp. Its nanoemulsion displayed lower antifungal activity, based on the changes in the chemical constituents due to homogenization by high-intensity ultrasound [80]. Further studies are necessary in order to explain the behavior of bioactive components during different encapsulation processes, especially the intensive-energy ones, and thereby evaluate the compatibility of the different encapsulation techniques for EOs.

## 8. Conclusions

Encapsulation represented an efficient approach to protect the EOs against environmental conditions that lead to oxidation or volatilization and reduced biological activities. Moreover, encapsulation solves the problem of EOs hydrophobicity and controls their release. Spray drying and emulsification are the most versatile and commercially available techniques used widely for EOs encapsulation. The encapsulated EOs showed enhanced antimicrobial, antifungal, antioxidant, antiviral, and insecticidal activities. The use of encapsulated EOs in food, cosmetics, and pharmaceutics could be an economic benefit and fulfill consumer concerns regarding safety. Energy-intensive techniques may negatively affect the structure-activity relationship of EOs bioactive components; therefore, further studies are necessary to find out the compatibility of encapsulation techniques for EOs.

## Funding sources

This research did not receive any specific grant from funding agencies in the public, commercial, or not-for-profit sectors.

## Conflict of interest

The authors declare that there is no conflict of interest.

## Author details

Hamdy A. Shaaban* and Amr Farouk
Department of Chemistry of Flavour and Aroma, National Research Center, Dokki, Giza, Egypt

*Address all correspondence to: hamdy.shaaban64@gmail.com

## References

[1] Seow YX, Yeo CR, Chung HL, Yuk HG. Plant essential oils as active antimicrobial agents. Critical Reviews in Food Science and Nutrition. 2014;**54**(5):625-644. DOI: 10.1080/10408398.2011.599504

[2] Ju J, Chen X, Xie Y, Yu H, Guo Y, Cheng Y, et al. Application of essential oil as a sustained release preparation in food packaging. Trends in Food Science and Technology. 2019;**92**(1800):22-32. DOI: 10.1016/j.tifs.2019.08.005

[3] Shetta A, Kegere J, Mamdouh W. Comparative study of encapsulated peppermint and green tea essential oils in chitosan nanoparticles: Encapsulation, thermal stability, in-vitro release, antioxidant, and antibacterial activities. International Journal of Biological Macromolecules. 2019;**126**:731-742. DOI: 10.1016/j.ijbiomac.2018.12.161

[4] Kavousi HR, Fathi M, Goli SAH. Novel cress seed mucilage and sodium caseinate microparticles for encapsulation of curcumin: An approach for controlled release. Food and Bioproducts Processing. 2018;**110**:126-135. DOI: 10.1016/j.fbp.2018.05.004

[5] Yildirim S, Röcker B, Pettersen MK, Nilsen-Nygaard J, Ayhan Z, Rutkaite R, et al. Active packaging applications for food. Comprehensive Reviews in Food Science and Food Safety. 2018;**17**:165-199

[6] Papadochristopoulos A, Kerry J, Fegan N, Burgess C, Duffy G. Natural anti-microbials for enhanced microbial safety and shelf-life of processed packaged meat. Foods. 2021;**10**:1598

[7] Liguori G, Sortino G, Gullo G, Inglese P. Effects of modified atmosphere packaging and chitosan treatment on quality and sensorial parameters of minimally processed cv. 'Italia' table grapes. Agronomy. 2021;**11**:328

[8] Salmieri S, Islam F, Khan RA, Hossain FM, Ibrahim H, Miao C, et al. Antimicrobial nanocompositefilms made of poly(Lactic acid)–cellulose nanocrystals (PLA–CNC) in food applications—Part B: Effect of oregano essential oilrelease on the inactivation of Listeria monocytogenes in mixed vegetables. Cellulose. 2014;**21**:4271-4285

[9] Dumitriu RP, Mitchell GR, Davis FJ, Vasile C. Functionalized coatings by electrospinning for antioxidant food packaging. Procedia Manufacturing. 2017;**12**:59-65

[10] Chen C, Xu Z, Ma Y, Liu J, Zhang Q, Tang Z, et al. Properties, vapour-phase antimicrobial and antioxidantactivities of active poly (Vinyl alcohol) packaging films incorporated with clove oil. Food Control. 2018;**88**:105-112

[11] Fang Z, Zhao Y, Warner RD, Johnson S. Active and intelligent packaging in meat industry. Trends in Food Science and Technology. 2017;**61**:60-71

[12] Kwon S-J, Chang Y, Han J. Oregano essential oil-based natural antimicrobial packaging film to inactivate Salmonella enteric and yeasts/molds in the atmosphere surrounding cherry tomatoes. Food Microbiology. 2017;**65**:114-121

[13] Munteanu SB, Vasile C. Vegetable additives in food packaging polymeric materials. Polymers. 2019;**12**:28

[14] Vasile C, Baican M. Progresses in food packaging, food quality, and

safety—Controlled-release antioxidant and/or antimicrobialpackaging. Molecules. 2021;**26**:1263

[15] Wan Z-L, Guo J, Yang X-Q. Plant protein-based delivery systems for bioactive ingredients in foods. Food and Function. 2015;**6**:2876-2889

[16] Taghavi T, Kim C, Rahemi A. Role of natural volatiles and essential oils in extending shelf life and controlling postharvest micro organisms of small fruits. Microorganisms. 2018;**6**:104

[17] Baptista-Silva S, Borges S, Ramos OL, Pintado M, Sarmento B. The progress of essential oils as potential therapeutic agents:A review. Journal of Essential Oil Research. 2020;**32**:279-295

[18] Sharifi-Rad J, Sureda A, Tenore GC, Daglia M, Sharifi-Rad M, Valussi M, et al. Biological activities of essential oils: From plant chemoecology to traditional healing systems. Molecules. 2017;**22**:70

[19] Dhifi W, Bellili S, Jazi S, Bahloul N, Mnif W. Essential oils' chemical characterization and investigation of some biological activities: A critical review. Medicines. 2016;**3**:25

[20] Banasaz S, Morozova K, Ferrentino G, Scampicchio M. Encapsulation of lipid-soluble bioactives by nanoemulsions. Molecules. 2020;**25**:3966

[21] Vuko E, Dunkić V, Ruščić M, Nazlić M, Mandić N, Soldo B, et al. Chemical composition and newbiological activities of essential oil and hydrosol of hypericum perforatum L. ssp. veronense (Schrank) H. Lindb. Plants. 2021;**10**:1014

[22] Sharmeen J, Mahomoodally F, Zengin G, Maggi F. Essential oils as natural sources of fragrance compounds for cosmeticsand cosmeceuticals. Molecules. 2021;**26**:666

[23] Bhavaniramya S, Vishnupriya S, Al-Aboody MS, Vijayakumar R, Baskaran D. Role of essential oils in food safety: Antimicrobialand antioxidant applications. Grain Oil Science and Technology. 2019;**2**:49-55

[24] Pandey AK, Kumar P, Singh P, Tripathi NN, Bajpai VK. Essential oils: Sources of antimicrobials and food preservatives. Frontiers in Microbiology. 2017;7:2161

[25] Vasile C, Sivertsvik M, Mitelut AC, Brebu MA, Stoleru E, Rosnes JT, et al. Comparative analysis of the composition and active property evaluation of certain essential oils to assess their potential applications in active food packaging. Materials. 2017;**10**:45

[26] Wu Z, Tan B, Liu Y, Dunn J, Guerola PM, Tortajada M, et al. Chemical composition and antioxidant properties of essential oils from peppermint, native spearmint and scotch spearmint. Molecules. 2019;**24**:2825

[27] Gutiérrez-Del-Río, I.; López-Ibáñez, S.; Magadán-Corpas, P.; Fernández-Calleja, L.; Pérez-Valero, Á.; Tuñón-Granda, M.; Miguélez,E.; Villar, C.; Lombó, F. Terpenoids and polyphenols as natural antioxidant agents in food preservation. Antioxidants 2021, 10,1264.

[28] Cardoso-Ugarte GA, Sosa-Morales ME. Essential oils from herbs and spices as natural antioxidants: Diversity of promisingfood applications in the past decade. Food Reviews International. 2021:1-31

[29] Amorati R, Foti MC, Valgimigli L. Antioxidant activity of essential oils. Journal of Agricultural and Food Chemistry. 2013;**61**:10835-10847

[30] Čabarkapa I, Čolović R, Đuragić O, Popović S, Kokić B, Milanov D, et al. Anti-biofilm activities of essential oils rich incarvacrol and thymol against Salmonella enteritidis. Biofouling. 2019;**35**:361-375

[31] Guimarães AC, Meireles LM, Lemos MF, Guimarães MCC, Endringer DC, Fronza M, et al. Antibacterial activityof terpenes and terpenoids present in essential oils. Molecules. 2019;**24**:2471

[32] Swamy MK, Akhtar MS, Sinniah UR. Antimicrobial properties of plant essential oils against human pathogens and theirmode of action: An updated review. Evidence-Based Complementary and Alternative Medicine. 2016;**2016**:3012462

[33] Mannu A, Melito S, Petretto GL, Manconi P, Pintore GM, Chessa M. Geographical variation of the chemical compositionin essential oils extracted from Sardinian Salvia verbenaca. Natural Product Research. 2020;7:1-4

[34] Sarmoum R, Haid S, Biche M, Djazouli Z, Zebib B, Merah O. Effect of salinity and water stress on the essential oil components of rosemary (rosmarinus officinalis L.). Agronomy. 2019;**9**:214

[35] A'cimovi'c M, Pezo L, Zeremski T, Lonˇcar B, Jeromela AM, Jeremic JS, et al. Weather conditions influence on hyssop essential oil quality. Processes. 2021;**9**:1152

[36] Abba BN, Ilagouma AT, Amadou I, Romane A. Chemical profiling, antioxidant and antibacterial activities of essential oil from englerastrumgracillimum Th. C. E. Fries Growing in Niger. Natural Product Communications. 2021;**16**:1934578X211002422

[37] Rao J, Chen B, Mc Clements DJ. Improving the efficacy of essential oils as antimicrobials in foods: Mechanisms of action. Annual Review of Food Science and Technology. 2019;**10**:365-387

[38] Kyriakoudi A, Spanidi E, Mourtzinos I, Gardikis K. Innovative delivery systems loaded with plant bioactive ingredients: Formulation approaches and applications. Plants. 2021;**10**:1238

[39] Rehman A, Jafari SM, Aadil RM, Assadpour E, Randhawa MA, Mahmood S. Development of active food packaging viaincorporation of biopolymeric nanocarriers containing essential oils. Trends in Food Science and Technology. 2020;**101**:106-121

[40] Ghareaghajlou N, Hallaj-Nezhadi S, Ghasempour Z. Red cabbage anthocyanins: Stability, extraction, biological activities andapplications in food systems. Food Chemistry. 2021;**365**:130482

[41] Zhuang S, Li Y, Jia S, Hong H, Liu Y, Luo Y. Effects of pomegranate peel extract on quality and microbiota composition ofbighead carp (Aristichthys nobilis) fillets during chilled storage. Food Microbiology. 2019;**82**:445-454

[42] Chambre DR, Moisa C, Lupitu A, Copolovici L, Pop G, Copolovici D-M. Chemical composition, antioxidant capacity, andthermal behavior of Satureja hortensis essential oil. Scitific Reports. 2020;**10**:21322

[43] Pavoni L, Perinelli DR, Bonacucina G, Cespi M, Palmieri GF. An overview of micro- and nanoemulsions as vehicles foressential oils: Formulation, preparation and stability. Nanomaterials. 2020;**10**:135

[44] Mucha W, Witkowska D. The applicability of essential oils in different

stages of production of animal-based foods. Molecules. 2021;**26**:3798

[45] Mohammadalinejhad S, Kurek M. Microencapsulation of anthocyanins—Critical review of techniques and wall materials. Applied Science. 2021;**11**:3936

[46] Ali A, Chen Y, Liu H, Yu L, Baloch Z, Khalid S, et al. Starch-based antimicrobial films functionalized bypomegranate peel. International Journal of Biological Macromolecules. 2019;**129**:1120-1126

[47] BeltránSanahuja A, Valdés García A. New trends in the use of volatile compounds in food packaging. Polymers. 2021;**13**:1053

[48] Pateiro M, Gómez B, Munekata P, Barba F, Putnik P, Kovaˇcevi'c D, et al. Nanoencapsulation of promising bioactivecompounds to improve their absorption, stability, functionality and the appearance of the final food products. Molecules. 2021;**26**:1547

[49] He L, Lan W, Ahmed S, Qin W, Liu Y. Electrospun polyvinyl alcohol film containing pomegranate peel extract and sodiumdehydroacetate for use as food packaging. Food Packag. Shelf Life. 2019;**22**:100390

[50] Serrano-Casas V, Pérez-Chabela ML, Cortés-Barberena E, Totosaus A. Improvement of lactic acid bacteria viability in acidconditions employing agroindustrial co-products as prebiotic on alginate ionotropic gel matrix co-encapsulation. Journal of Functional Foods. 2017;**38**:293-297

[51] Lin L, Zhu Y, Cui H. Electrospun thyme essential oil/gelatin nanofibers for active packaging against Campylobacter jejuni inchicken. LWT. 2018;**97**:711-718

[52] FonsecaLM, CruxenCEDS, BruniGP, Fiorentini ÂM, Zavareze EDR, Lim L-T, et al. Development ofantimicrobial and antioxidant electrospun soluble potato starch nanofibers loaded with carvacrol. International Journal of Biological Macromolecules. 2019;**139**:1182-1190

[53] López de Dicastillo C, López-Carballo G, Gavara R, Muriel Galet V, Guarda A, Galotto MJ. Improving polyphenolicthermal stability of Aristotelia Chilensis fruit extract by encapsulation within electrospun cyclodextrin capsules. Journal of Food Processing and Preservation. 2019;**43**:e14044

[54] Saucedo-Zuñiga J, Sánchez-Valdes S, Ramírez-Vargas E, Guillen L, Ramos-Devalle L, Graciano-Verdugo A, et al. Controlled release of essential oils using laminarnanoclay and porous halloysite/essential oil composites in a multilayer film reservoir. Microporous Mesoporous Materials. 2021;**316**:110882

[55] BastosLPH, VicenteJ, dosSantosCHC, de Carvalho MG, Garcia-Rojas EE. Encapsulation of black pepper (Piper nigrumL.) essential oil with gelatin and sodium alginate by complex coacervation. Food Hydrocoll. 2020;**102**:105605

[56] Rosli NA, Hasham R, Aziz AA. Design and physicochemical evaluation of lipid encapsulated Zingiber zerumbet oil by d-optimal mixture design. JurnalTeknologi. 2018;**80**(3):105-113. DOI: 10.11113/jt.v80.11268

[57] Chung SK, Seo JY, Lim JH, Park HH, Yea MJ, Park HJ. Microencapsulation of essential oil for insect repellent in food packaging system. Journal of Food Science. 2013;**78**(5):1-6. DOI: 10.1111/1750-3841.12111

[58] Cortés-Camargo S, Acuña-Avila PE, Rodríguez-Huezo ME, Román-Guerrero A, Varela-Guerrero V, Pérez-Alonso C. Effect of chia mucilage addition on oxidation and release kinetics of lemon essential oil microencapsulated using mesquite gum—Chia mucilage mixtures. Food Research International. 2019;**116**:1010-1019. DOI: 10.1016/j.foodres.2018.09.040

[59] Ngamekaue N, Chitprasert P. Effects of beeswax-carboxymethyl cellulose composite coating on shelf-life stability and intestinal delivery of holy basil essential oil-loaded gelatin microcapsules. International Journal of Biological Macromolecules. 2019;**135**:1088-1097. DOI: 10.1016/j.ijbiomac.2019.06.002

[60] ArdiansyahRukmana RAF, Warsito W, Rahman MF. Microencapsulation of Kaffir Lime Oil (citrus hystrix d. c.) from peels and twigs oil fraction and its activity as antioxidant. In: Proceedings of International Conference of Essential Oil (ICEO 2017). Indonesia: Brawijaya University; 2017. pp. 112-118

[61] Siow L, Ong C. Effect of pH on garlic oil encapsulation by complex coacervation. Journal of Food Processing & Technology. 2012;**04**(01):1-5. DOI: 10.4172/2157-7110.1000199

[62] Vishwakarma GS, Gautam N, Babu JN, Mittal S, Jaitak V. Polymeric encapsulates of essential oils and their constituents: A review of preparation techniques, characterization, and sustainable release mechanisms. Polymer Reviews. 2016;**56**(4):668-701. DOI: 10.1080/15583724.2015.1123725

[63] Xiao Z, Liu W, Zhu G, Zhou R, Niu Y. Production and characterization of multinuclear microcapsules encapsulating lavender oil by complex coacervation. Flavour and Fragrance Journal. 2014;**29**(3):166-172. DOI: 10.1002/ffj.3192

[64] Wang B, Akanbi TO, Agyei D, Holland BJ, Barrow CJ. Coacervation technique as an encapsulation and delivery tool for hydrophobic bio-functional compounds. In: Role of Materials Science in Food Bioengineering. London, United Kingdom: Academic Press; 2018. pp. 235-261

[65] Khoshakhlagh K, Koocheki A, Mohebbi M, Allafchian A. Development and characterization of electrosprayed *Alyssum homolocarpum* seed gum nanoparticles for encapsulation of D-limonene. Journal of Colloid and Interface Science. 2017;**490**:562-575. DOI: 10.1016/j.jcis.2016.11.067

[66] Tolun A, Altintas Z, Artik N. Microencapsulation of grape polyphenols using maltodextrin and gum Arabic as two alternative coating materials: Development and characterization. Journal of Biotechnology. 2016;**239**:23-33. DOI: 10.1016/j.jbiotec.2016.10.001

[67] Fraj J, Petrović L, Đekić L, Budinčić JM, Bučko S, Katona J. Encapsulation and release of vitamin C in double W/O/W emulsions followed by complex coacervation in gelatin-sodium caseinate system. Journal of Food Engineering. 2021;**292**(2020):110353. DOI: 10.1016/j.jfoodeng.2020.110353

[68] Yada RY, Buck N, Canady R, DeMerlis C, Duncan T, Janer G, et al. Engineered nanoscale food ingredients: Evaluation of current knowledge on material characteristics relevant to uptake from thegastrointestinal tract. Comprehensive Reviews in Food Science and Food Safety. 2014;**13**:730-744

[69] Joye IJ, McClements DJ. Biopolymer-based nanoparticles and microparticles:

Fabrication, characterization, and application. Current Opinion in Colloid Interface Science. 2014;**19**:417-427

[70] Zhang Y, Hsu BY, Ren C, Li X, Wang J. Silica-based nanocapsules: Synthesis, structure control and biomedical applications. Chemical Society Reviews. 2014;**44**:315-335

[71] Shishir, M.R.I.; Xie, L.; Sun, C.; Zheng, X.; Chen,W. Advances in micro and nano-encapsulation of bioactive compounds usingbiopolymer and lipid-based transporters. Trends Food Science and Technology. 2018, 78, 34-60.

[72] Shin GH, Kim JT, Park HJ. Recent developments in nanoformulations of lipophilic functional foods. Trends Food Science and Technology. 2015;**46**:144-157

[73] Prakash A, Baskaran R, Paramasivam N, Vadivel V. Essential oil based nanoemulsions to improve the microbial quality ofminimally processed fruits and vegetables: A review. Food Research International. 2018;**111**:509-523

[74] McClements DJ, Li Y. Structured emulsion-based delivery systems: Controlling the digestion and release of lipophilic foodcomponents. Advances in Colloid and Interface Science. 2010;**159**:213-228

[75] de Souza Simões L, Madalena DA, Pinheiro AC, Teixeira JA, Vicente AA, Ramos OL. Micro-and nano bio-based deliverysystems for food applications: In vitro behavior. Advances in Colloid and Interface Science. 2017;**243**:23-45

[76] Jia Z, Dumont M-J, Orsat V. Encapsulation of phenolic compounds present in plants using protein matrices. Food Bioscience. 2016;**15**:87-104

[77] Chang Y, McLandsborough L, McClements DJ. Physical properties andantimicrobial efficacy of thymeoil nanoemulsions: Influence of ripening inhibitors. Journal of Agriculture and Food Chemistry. 2012;**60**:12056-12063. DOI: 10.1021/jf304045a

[78] Ali H, Al-Khalifa AR, Aouf A, Boukhebti H, Farouk A. The effect of nanoencapsulation andnanoemulsion on the chemical structure and biological activity of Algerian *Origanum glandulosum*. Scientific Reports. 2020;**10**:2812. DOI: 10.1038/s41598-020-59686-w

[79] Aouf A, Ali H, Al-Khalifa AR, Mahmoud KF. Farouk, A. Influence of Nanoencapsulation Using High-Pressure Homogenization on the Volatile Constituents and Anticancer and Antioxidant Activities of Algerian *Saccocalyxsatureioides* Coss. et Durieu. Molecules. 2020;**25**:4756. DOI: 10.3390/molecules25204756

[80] Farouk A, Hathout A, Amer M, Hussain O, Fouzy A. The impact of nanoencapsulation on volatile constituents of Citrus sinesis L. essential oil and their antifungal activity. Egyptian Journal of Chemistry. 2022;**65**(3):1-2. DOI: 10.21608/ejchem.2021.95579.4485

Section 3

# Essential Oils with Pharmacological Properties

Chapter 6

# Biological Application of Essential Oils and Essential Oils Components in Terms of Antioxidant Activity and Inhibition of Cholinesterase Enzymes

*Mejra Bektašević and Olivera Politeo*

## Abstract

This chapter will be described oxidative stress related to modern age illness as well as biological activity of essential oils and essential oil components in terms of their antioxidant activity. The importance of essential oils and their constituents in terms of protecting lipids and proteins from oxidation will also be explained. Alzheimer's disease as a disease related to oxidative stress and strategies in their treatment by using essential oil components as cholinesterase inhibitors will also be described. As case studies will be pointed out medicinal plants, endemic *Saturejasubspicata* L., and widely used *Menthapulegium* L. growing in Bosnia and Herzegovina.

**Keywords:** oxidative stress, essential oils, biological activity, antioxidants, Alzheimer's disease, cholinesterase inhibitors, *Saturejasubspicata*, *Menthapulegium*

## 1. Introduction

Under normal physiological conditions, the production of harmful reactive species caused by oxidative processes and antioxidant defense are in balance. If the reactive oxygen species and other species production exceed the antioxidant capacity of a living system, reactive oxygen and nitrogen species (ROS and RNS) may react with macromolecules, causing structural and/or functional damage to cellular enzymes and genetic material. An excess of reactive species and damage caused by their action is called oxidative stress.

In a state of oxidative stress, an excess of ROS and RNS may damage lipids, proteins, carbohydrates, and nucleic acids. Free radicals attack unsaturated fatty acids in biological membranes causing lipid peroxidation. Lipid peroxidation is an enzymatic reaction catalyzed by the enzyme lipoxygenase [1]. This enzyme is found in the erythrocytes and leukocytes of animals, as well as in many plant organisms. Its substrate is linoleic and linolenic acid in plants, and arachidonic acid in animals, while oleic acid is not oxidized. Lipid peroxidation results in decreased membrane fluidity, loss of enzymes and receptor activity, damage to membrane proteins and other macromolecules, which leads to apoptosis [2].

Oxidative modification of proteins, reversible and irreversible, occurs during redox signaling and other cellular processes. It also occurs as a result of oxidative stress. Exposure of proteins to hydroxyl $OH^{\bullet}$ and/or superoxide radicals $O_2^{\bullet-}$ leads to their structural modifications. Modified proteins may further undergo spontaneous fragmentation and cross-linking or show a significant increase in proteolysis. An oxidative attack of a polypeptide backbone is usually initiated by hydroxyl $OH^{\bullet}$. By an experimental generation of radicals, using water radiolysis or decomposing hydrogen peroxide $H_2O_2$ in a metal-catalyzed reaction - and in the interaction with lipids - alkyl, alkoxy, and alkylperoxyl radical intermediates can be formed, which affect peptide bond cleavage in several ways.

Tryptophan, histidine, and cysteine are the most sensitive to reactive oxygen species. In addition to fragmentation, oxidation of the amino acid residues of lysine, arginine, proline, and threonine increases carbonyl concentration, so the presence of carbonyl groups can be used as an indicator of protein oxidation.

Oxidative modification of proteins also occurs in reaction with aldehydes, which are formed during lipid peroxidation process. End products of lipid peroxidation, such as malondialdehyde (MDA) and 4-hydroxy-2-nonenal (HNE), as well as oxidation products of polyunsaturated fatty acids cause oxidative damage to proteins [3].

Oxidative modification of proteins is present in diseases and changes associated with the aging process, such as atherosclerosis, tumors, neurodegenerative diseases, and aging. Protein carbonylation occurs with a large number of modifications and is a marker of oxidative stress. During the first two-thirds of life, the level of protein carbonylation slowly increases, while its level rises sharply in the last third. Protein carbonylation negatively affects the functions of proteins themselves, which suggests that this modification may be one of the causes of the aforementioned undesirable processes [4].

## 2. Oxidation and food

Apart from the living organisms, the oxidation process occupies an important place in the food, pharmaceutical, and cosmetic industries. It includes the oxidation of protein molecules, vitamins, but above all, the oxidation of lipid molecules [5].

Oxidation of lipid molecules is a major problem in the food industry, as it leads to changes in the organoleptic properties of food, a decrease in its nutritional value, as well as the formation of radical components that can endanger consumers' health.

Lipid oxidation in food implies a whole range of chemical changes that result from the reaction of lipids with oxygen. Triacylglycerols and phospholipids are hardly volatile molecules and do not directly affect the aroma of the product. During lipids oxidation from fatty acids, volatile compounds have formed that lead to an undesirable aroma of products known as rancidity [6].

Polyunsaturated fatty acids oxidize much faster than monounsaturated or saturated ones. The rate of lipid oxidation is influenced by the number and position of double bonds [1]. The methylene group ($-CH_2-$) located between the two double bonds is very susceptible to oxidation. Linoleic acid is subject to oxidation, as it has a methylene group between two double bonds, at position 11. Its oxidation produces two hydroperoxides. The main secondary product of linoleic acid autooxidation is hexanal. Lipid autooxidation is an autocatalytic reaction, which means that it progresses over time due to the formation of products that catalyze the reaction themselves.

**Initiation:** $LH \rightarrow L^{\bullet} + H^{\bullet}$

**Propagation:** $L^{\bullet} + O_2 \rightarrow LOO^{\bullet}$
$LOO^{\bullet} + LH \rightarrow LOOH + L^{\bullet}$
$L^{\bullet} + O_2 \rightarrow LOO^{\bullet}$

**Termination:** $2LOO^{\bullet}$
$LOO^{\bullet} + L^{\bullet}$
$L^{\bullet} + L^{\bullet}$

**Figure 1.**
*The lipid oxidation phases [7].*

Lipid peroxidation includes three phases: initiation, propagation, and termination (**Figure 1**). From the peroxides formed at the beginning, secondary oxidation products are formed: aldehydes, ketones, epoxides, and other compounds, which also have negative biological effects, such as loss of essential amino acids and lipid-soluble vitamins [7].

In the first phase, oxygen from the air attacks unsaturated fatty acids (LH), creating free radicals of fatty acids (peroxy $LO_2^{\bullet}$, alkoxyl $LO^{\bullet}$, or alkyl radicals $L^{\bullet}$). In the second phase of the reaction, hydroperoxides (LOOH) and free peroxide radicals ($LOO^{\bullet}$) are formed from free radicals by binding oxygen to free fatty acid radicals.

Hydroperoxides (primary oxidation products) are labile, so they are further decomposed into free radicals and decomposed oxidation products. These degradation products of oxidation (secondary oxidation products) are carbonyl compounds (aldehydes and ketones), fatty acids, alcohols, epoxides, etc., some of which give off an unpleasant, rancid odor characteristic of oxidized fat.

Lipid autooxidation is often initiated by free radicals from an unknown source. It is accelerated by rising temperatures, light and the presence of trace metals. Reductive forms of transition metals are more efficient in the hydrogen peroxide decomposition, so reductive components such as superoxide anion ($O_2^{\bullet-}$) and ascorbic acid further promote lipid oxidation. Redox cycling of iron in the presence of superoxide anions in lipid oxidation is known as the Haber-Weiss reaction, while the second step of this reaction is known as the Fenton reaction:

$$Fe^{3+} + O_2^{\bullet-} \rightarrow Fe^{2+} + O_2.$$

$$Fe^{2+} + H_2O_2 \rightarrow Fe^{3+} + OH^{\bullet} + OH^{-}.$$

The resulting hydroxyl radicals ($OH^{\bullet}$) are the most reactive ROS species.

Ascorbic acid can also participate in the Haber-Weiss type reaction, but unlike superoxide anions, ascorbic acid may also act as an antioxidant at higher concentrations.

The control of the level of free radicals, prooxidants, and oxidation intermediates is used to protect the lipid components of food from oxidation. Free radical scavengers (FRS) inhibit lipid oxidation by reacting faster than unsaturated fatty acids with free radicals. They can react with peroxyl($LOO^{\bullet}$) or alkoxyl($LO^{\bullet}$) radicals in the following reaction:

$$LOO^{\bullet} \text{ or } LO^{\bullet} + FRS \rightarrow LOOH \text{ or } LOH + FRS^{\bullet}.$$

Phenolic components are known to be good free radicals scavengers, as they donate a hydrogen atom, and the resulting radical has low energy due to its delocalization in the structure of phenol ring (**Figure 2**) [6].

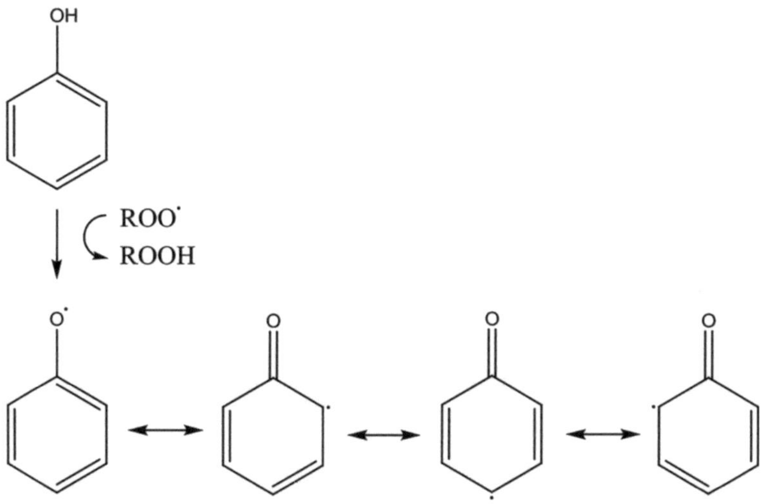


**Figure 2.**
*Delocalization of phenol radical [6].*

The most commonly used synthetic antioxidants are substituted monophenolic compounds, such as 2,6-di-*tert*-butyl-4-hydroxytoluene (BHT), 2-*tert*-butyl-4-hydroxyanisole and 3-*tert*-butyl-4-hydroxyanisole (2- and 3-BHA), propyl gallate(PG) and *tert*-butyl hydroquinone (TBHQ). The addition of the antioxidant BHA prolongs the stability time of lipid-based foods (e.g., butter, fat, meat, dairy, vegetable oils) by a few months to a few years. BHT is less effective than BHA because two tertiary butyl groups sterically affect the radical reaction. PG is poorly soluble in water. It is less commonly used in the food industry because it binds $Fe^{3+}$ ions and reduces them to their $Fe^{2+}$ form. When this antioxidant is used in food, it must be combined with chelating agents (such as citrate) to prevent this phenomenon. TBHQ is one of the best antioxidants added to oil intended for frying. Unlike PG, it does not complex with iron and copper ions. According to European rules, the permitted amount of BHT, BHA, and PG in food is 100 μg/g of lipids [8]. BHA and BHT are very effective in their role, but are easily volatile and thermolabile, which makes their use limited [9]. Some studies have shown that the use of some antioxidants of synthetic origin has negative effects on human health due to the promotion of carcinogenesis [10, 11].

For these reasons, there is a tendency to replace synthetic antioxidants, where possible, with non-toxic antioxidants of natural origin. More recently, essential oils have also been used as a substitute for synthetic antioxidants, in those food canning sectors where their use does not adversely affect product flavor [12].

## 3. Antioxidant activity of essential oil components

In addition to oxidative damage and death of cells, tissue damage and various pathological conditions may be the consequence of oxidative stress. Numerous forms of malignant disease are thought to be the result of oxidative DNA damage and the resulting mutations. The negative impact of free radicals is believed to lead to various autoimmune diseases, diabetes, rheumatic diseases, cardiovascular disease and heart attack, kidney disease, infectious diseases, neurodegenerative diseases

(Alzheimer's disease), etc. The aging process itself is described as the process of accumulation of numerous oxidative damage accumulated over time.

Given that the oxidative stress is associated with the etiology and pathogenesis of many diseases, it is believed that eliminating the causes of oxidative stress may prevent or delay the occurrence of pathological changes and reduce the occurrence of diseases. Numerous studies show that regular intake of fruits, vegetables, grains, and beverages have a positive effect on diseases that are mediated by the activity of free radicals. Therefore, natural antioxidants – alone or in the form of extracts – may be useful in the treatment of such diseases. Thus, the reason for the great interest in researching the antioxidant activity of aromatic, medicinal, and edible plants [13].

In situations of disturbed homeostasis, as well as in the prevention of disease development, the intake of antioxidants in food may be of great importance. In this regard, essential oils, plant extracts, or their individual components with good antioxidant activity may be used. From a chemical point of view, essential oils are complex mixtures of a large number of compounds, which makes their activity difficult to test.

With the exception of some phenolic components, whose antimicrobial and antioxidant activity is well known, such data are not available for most other components of essential oils. Numerous papers on essential oils mention synergism, antagonism, additivity, but such claims are rarely accompanied by experimental confirmation [12].

A study by Ruberto and Baratta [12] examined the antioxidant activity of 100 pure compounds, common constituents of essential oils, using two methods. Of the thirteen non-oxygenated monoterpenes, terpinolene, α-terpinene, γ-terpinene, and sabinen showed very high activity. The activity of α-terpinene and γ-terpinene was similar to that shown by α-tocopherol. An active methylene group is thought to contribute to this activity of the aforementioned compounds. Of the 34 oxygenated monoterpenes tested, thymol and carvacrol showed activity as did α-tocopherol. It is known that thymol and carvacrol contribute the most to the antioxidant activity of essential oils that contain them. Alcohols were the most active in this class of compounds, with the exception of linalool, which showed prooxidative activity. Ketones showed lower activity. Non-oxygenated sesquiterpenes were not active, while oxygenated sesquiterpenes showed activity similar to that of oxygenated monoterpenes. Germacron, a cyclic ketone, showed slightly more pronounced activity, while nerolidol showed prooxidative activity. Phenols, benzene derivatives, have shown the best results. They are more effective in preventing the formation of primary oxidation products, as opposed to preventing the formation of secondary oxidation products. Non-terpene compounds, which are present in essential compounds in a smaller amount, showed weak antioxidant activity – just like non-oxygenated sesquiterpenes [12].

More recently, essential oils have also been used as a substitute for synthetic antioxidants, in those food preservation sectors where their use does not adversely affect product flavor [12].

Due to their specific chemical structure, plant phenolic compounds may act as strong antioxidants, due to their ability to interrupt chain reactions by donating hydrogen atom or electron to a free radical, while taking on a stable non-reactive conformation. However, their activity depends on a number of factors: degree of hydroxylation, polarity, solubility, reducing potential, stability of the resulting radical, etc. Hydroxycinnamic acids, the components of essential oils, show stronger activity compared to hydroxybenzoic acids because they donate hydrogen atoms more easily [14]. Polyphenols are proven to have a positive effect on cognitive abilities and neurodegenerative changes caused by aging [15].

Currently, there is a disparity in knowledge about the *in vivo* and *in vitro* effects of polyphenols as antioxidants [16]. Due to the lack of knowledge regarding the safety of higher doses intake, it is believed that the level of polyphenols, which are entered into the human body, should not exceed that in which they are otherwise found in food [17].

## 4. Neurotransmitter acetylcholine and cholinesterase inhibitors

The neurotransmitter acetylcholine (ACh) is present in the nervous system, where it enables cerebral-cortical activity and development, control of cerebral blood flow, control of sleep–wake cycles, as well as learning and memory processes (**Figure 3**). The enzyme cholineacetyltransferase (ChAT) catalyzes the production of acetylcholine (ACh) in cholinergic neurons, from choline and acetyl coenzyme A.

Releasing acetylcholine from the synaptic vesicle of the presynaptic membrane into the synaptic cleft, it binds to cholinergic receptors (nicotinic and muscarinic receptors) on the postsynaptic membrane of the cholinergic synapse or on muscle cells. This triggers a series of processes that result in membrane depolarization and further signal transmission [18].

ACh hydrolysis controls the transmission of nerve impulses at the cholinergic synapses of the central and peripheral nervous systems. The degradation of acetylcholine in the synaptic cleft by acetylcholinesterase (AChE) establishes the polarization of the postsynaptic membrane and impulse transmission ceases.

Two types of ChE are currently known: acetylcholinesterase (AChE) and butyrylcholinesterase (BChE). AChE is also called "true cholinesterase", while BChE is also known as "pseudocholinesterase" because it hydrolyzes many choline esters and other non-choline esters (butyrylcholine, succinylcholine, acetylcholine, acetylsalicylic acid, cocaine, and heroin).

Inhibition of AChE prevents the hydrolysis of ACh, thus prolonging its activity in the transmission of nerve impulses. This concept is applied in the treatment of diseases characterized by low ACh levels and is also being studied in toxicology because of health conditions and deaths caused by increased cholinergic stimulation [19].

Alzheimer's disease (AD) is the most common neurodegenerative disorder and the cause of dementia in the elderly population. It affects about 2% of the population in industrialized countries. AD is characterized by the formation of neuritic plaques; extracellular accumulations of fibrils and amyloid-β-peptides, as well as neurofibrillary tangles; intracellular accumulations of τ-protein, in regions of the brain responsible for learning, memory, and emotional behavior. These changes cause neuronal degeneration, loss of choline acetyltransferase (ChAT) and acetylcholinesterase (AChE), which is manifested in the loss of neurotransmitters and other neuromodulators, and

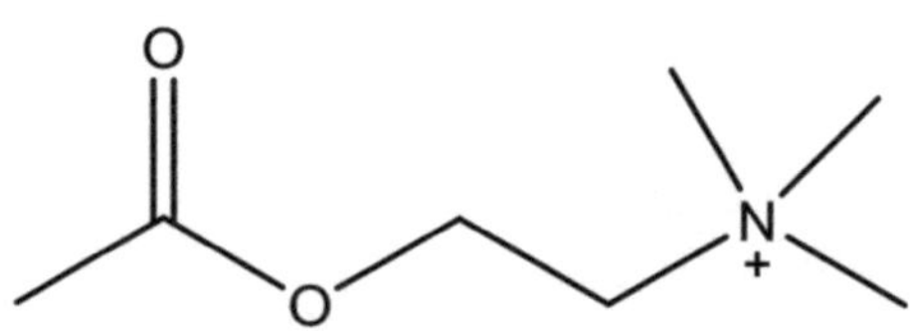


**Figure 3.**
*Structural formula of neurotransmitter acetylcholine (ACh).*

the disabling of synaptic transmission [18]. Currently, the treatment of this disease is limited to the treatment of symptoms of the disease, for which cholinesterase inhibitors (ChE) are used.

ChE inhibitors may be reversible, those which are bound by noncovalent interactions, or irreversible, which covalently bind to the serine of the catalytic triad. Reversible inhibitors bind to the active site, peripheral site or both, and the inhibition occurs as a result of conformational changes of the enzyme, electrostatic interactions of the inhibitor and the cationic part of the substrate, and steric and/or electrostatic interferences with the substrate entry into the active enzyme center.

A feature of the structure of good cholinesterase inhibitors is the presence of a positive charge and/or aromatic or hydrophobic substituents that facilitate the entry and placement of inhibitors in the active site of the enzyme [18].

Synthetic AChE inhibitors such as physostigmine, tacrine, and donepezil cause side effects such as hepatotoxicity and gastrointestinal disorders. Irreversible inhibitors may cause serious consequences and even death, as is the case with sarin, a poison gas, so reversible inhibitors are preferred in this regard [20].

### 4.1 Essential oil components as cholinesterase inhibitors

Bioactive substances from fruits, vegetables, and medicinal plants play a major role in slowing down many pathogeneses and neurodegenerative disorders, such as Alzheimer's disease. In addition to alkaloids, food rich in phytochemicals contains terpenes and polyphenols, which can be good cholinesterase inhibitors, alone or in synergy with each other [20].

Donepezil, rivastigmine, and galantamine are currently used to treat AD symptoms, such as cognitive dysfunction and memory impairment [21]. The aforementioned galantamine is a reversible inhibitor of AChE, which has been used since 2007 in the treatment of mild to moderate AD. It shows good pharmacological and pharmacokinetic properties, as well as a small number of side effects [22]. The use of most of the ChE inhibitors tested so far has been accompanied by side effects such as fatigue, sleep disorders, cardiorespiratory, gastrointestinal disorders, and low bioavailability. This was an incentive for further research with the aim of finding new ChE inhibitors of natural origin, with greater efficiency and bioavailability, as well as with fewer side effects [23].

Essential oils contain a number of bioactive components; terpenes, terpenoids, phenylpropanoid and other compounds, so a large number of them have been tested in terms of their ability to inhibit ChE. The results showed that some of the tested oils have a good ability to inhibit ChE. Comparing the results of different studies, it was noticed that some essential oils of similar composition have different abilities to inhibit ChE. The differences in the mentioned results may be attributed to the synergistic or antagonistic effect between the individual components of the essential oil. To investigate these effects, a number of studies have been conducted to identify and isolate individual constituents of essential oils with a significant ability to inhibit ChE [24].

The majority of the data obtained thus far in the research pertains to the study of the ability of smaller individual components of essential oils to inhibit AChE, while a few pertain to the study of BChE inhibition. However, given the role of BChE inhibition in the treatment of AD in the later stages of the disease, the interest in testing BChE inhibition has increased [24]. In terms of ChE inhibition, $IC_{50}$ values are impacted by the enzyme concentration, inhibitors, and substrates, as well as other experimental conditions, making it difficult to compare the results obtained by different studies. It is important to standardize the protocols used in testing AChE and BChE inhibitors, so as to be able to detect them [25].

When it comes to the studies of the ability to inhibit ChE, most of these refer to the study of monoterpenes [24]. Of monoterpenes, 1,8-cineole and α-pinene are the most effective in inhibiting AChE. In addition to these two, the ability to inhibit AChE is shown by δ-2-carene (2-carene), δ-3-carene (3-carene), and mirtenal [18, 24], as well as geraniol, α-caryophyllene, and limonene [21]. Carvone also showed good AChE inhibitory activity [19].

Monoterpene carvacrol and its isomer thymol showed significant AChE inhibitory activity, with carvacrol activity being ten times higher, which indicates the importance of the hydroxyl group position for AChE inhibitory activity [26].

Among the monoterpenes with the *p*-menthane skeleton, pulegone was the most successful in terms of AChE inhibition [19]. Monoterpene camphor, bornyl acetate, carvone, β-pinene, fenchol, and fenchone show poorer ability to inhibit AChE [19].

Some studies show the existence of a synergistic effect of monoterpenes, especially between 1,8-cineole and α-pinene [19]. A synergistic effect is also present between the enantiomers of α-pinene and β-pinene (α-*S*-pinene: β-*S*-pinene, and α-*R*-pinene: β-*S*-pinene) in the mixture, at a ratio of 3:2 [27].

One of the ways in which terpenes inhibit AChE is through a hydrophobic ligand. The hydrophobic active site of AChE is the site where hydrophobic interactions take place, and terpene compounds, built from the skeletons of carbon and hydrogen atoms, thus contribute to the inhibition of cholinesterases [21].

Due to the differences in terpene compounds structure, it is difficult to determine the relationship between their structure and activity. When it comes to monoterpenes with a *p*-menthane skeleton, it has been established that ketone monoterpenes are better AChE inhibitors than the corresponding hydrocarbons and alcohols [19]. In the case of bicyclic monoterpenes, bicyclic hydrocarbons have a greater ability of inhibition than bicyclic alcohols and ketones. The position of the double bond increases the ability of inhibition, so 3-carene has a greater ability of inhibition than 2-carene [28].

Monoterpenes are much better inhibitors of AChE than BChE. Due to their low molecular weight, monoterpenes are more likely to inhibit ChE exerting steric or allosteric effects, whereby BChE does not affect the substrate's access to the enzyme site [18].

In a study examining 21 monoterpenes in terms of the ability to inhibit BChE, only 3carene showed BChE inhibiting ability ($IC_{50}$ = 2000 μM) [29]. Monoterpenes α-pinene, 1,8-cineole, 1,8-cineole, linalool, terpinen-4-ol, linalyl acetate, thymol, γ-terpinene, and phenylpropanoid eugenol have shown good to moderate BChE inhibitory potential ($IC_{50}$ = 0,1 to 1,0 mM) in various studies [24].

Of the flavonoids, flavones and isoflavones show the best activity, while xanthones and monoterpenes show weaker activity in the inhibition of cholinesterases (**Figure 4**) [18].

The most frequently studied sesquiterpenes for AChE inhibition are β-caryophyllene and α-humulene. In doing so, β-caryophyllene had a good ability to inhibit, in contrast to α-humulene (α-caryophyllene), which showed a weak ability to inhibit AChE [24]. In several studies, β-caryophyllene also showed good to moderate BChE inhibitory potential ($IC_{50}$ = 0,1 to 1,0 mM) [24].

Diterpenes inhibit ChE at lower concentrations than monoterpenes, which indicates the importance of molecule size. Dihydrotanshinone and cryptotanshinone are non-competitive ChE inhibitors. Of triterpenes and steroids, ursolic acid, taraxerol, leucisterol, and oleanolic acid show ChE inhibitory activity [18].

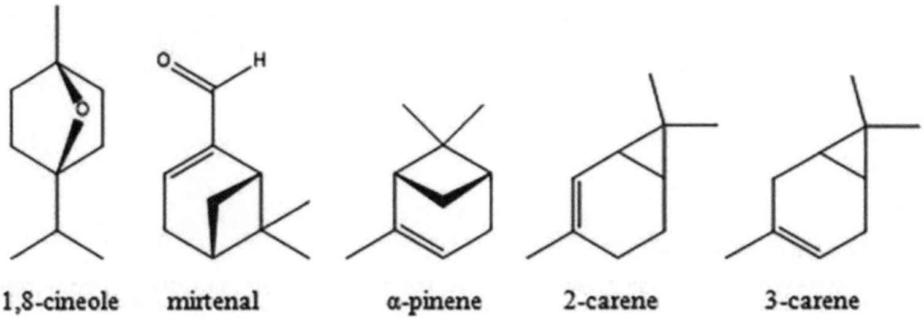


**Figure 4.**
*Some monoterpenes with cholinesterase inhibition activity.*

Given that BChE has a regulatory role in ACh hydrolysis, therapeutics that would inhibit both ChEs could exert additional positive effects in the treatment of AD, compared to inhibitors that inhibit only AChE. Thus, rivastigmine, which inhibits both ChEs, is very successful in the AD treatment. To date, there is no evidence that BChE inhibitors are more effective in reducing AD symptoms than AchE inhibitors [18].

In traditional medicine, many herbs are used in the treatment of cognitive disorders, including neurodegenerative diseases. The ethnopharmacological approach, testing of biological activity and isolation enabled the identification of potential AChE inhibitors of plant origin. Multifunctional compounds with several complementary biological functions are of particular interest. Plant extracts are the main sources of new compounds, AChE inhibitors [21]. In this regard, polyphenols are particularly interesting due to their positive effect on human health [20].

Many phytochemicals are bioactive compounds, some of which show ChE inhibitory activity and represent a model for the development of new drugs, ChE inhibitors. As terpenes and terpenoids have shown relatively weak inhibitory capacity in studies published so far, it is necessary to develop analogues with an improved efficiency [21].

Given the above, numerous plant extracts and essential oils, as well as their components, have been studied in terms of ChE inhibitory activity [18, 19, 20].

## 5. The case study of *Satureja subspicata* and *Mentha pulegium* growing in Bosnia and Herzegovina

*Satureja subspicata* (mountain savory) is a rare, endemic Dinaric species distributed in the eastern Mediterranean area [30]. *Satureja subspicata* (**Figure 5**) has long been utilized in Bosnia-Herzegovina's traditional medicine to treat leukemia and lymph node disorders. The herbal medicines of *Satureja subspicata* have been shown to be effective for a variety of cardiovascular diseases, especially arrhythmia, atrial fibrillation, and vascular diseases [31].

The essential oils obtained from various *Satureja* species have certain biological properties, such as antimicrobial [32, 33], antioxidant [33, 34], antiviral [35], antispasmodic and antidiarrheal [36, 37], anti-inflammatory and antinociceptive activities [38]. Carvacrol, thymol, β-caryophyllene, γ-terpinene, *p*-cymene, and linalool, all common compounds found in *Satureja* essential oil, have been shown to have strong antioxidant activity [39].

**Figure 5.**
Satureja subspicata *L. (mountain savory).*

Thirty-four (34) volatile compounds (98.0% of the total oil) in *Satureja subspicata* essential oil from Bosnia-Herzegovina were identified using GC/MS and GC/FID. The main classes of essential oil constituents were non-oxygenated monoterpenes (46.6%) and non-oxygenated sesquiterpenes (34.8%), followed by oxygenated monoterpenes (10.4%) and oxygenated sesquiterpenes (6.2%). The sesquiterpene ß-caryophyllene (14%) and non-oxygenated monoterpene *cis*-ß-ocimene (12.1%), as well as α-pinene (10.2%), were the main essential oil components. Other quantitatively important compounds were *trans*-ß-ocimene (8.8%), germacrene D (7.1%), caryophyllene oxide (6.2%), and myrtenol (6.1%) [40].

*Mentha pulegium* L. (pennyroyal) can be found in the area of Europe and Mediterranean [41]. In traditional medicine of Bosnia and Herzegovina, this plant is used for the treatment of nervous system disorders [31]. *Mentha pulegium* (**Figure 6**) stimulates digestive juices and helps with bloating and cramps. It is used for headaches and mild respiratory infections; it is a strong stimulant for the muscles of the uterus and can be used externally to relieve rheumatic problems, including gout [42].

Medicinal properties of *M. pulegium* are attributed to the monoterpenes present in the essential oil as well as polyphenol derivatives [43]. Essential oil of *M. pulegium* shows antifungal, insecticidal, antiparasitic, spasmolytic, and antioxidant properties [44]. Because of the mint-like odor, *M. pulegium* essential oil has a wide application; it is a constituent of foods and fragrances [45]. Pulegone, piperitone or piperitenone have been identified as dominant *M. pulegium* oil components [46]. Toxic effects of *M. pulegium* essential oil are mainly due to its main component pulegone. Reports suggest that the ingestion of up to 10 mL of *Mentha pulegium* oil causes gastritis and mild central nervous system toxicity without hepatorenal damage. The fatalities resulting from the ingestion of 15 to 30 mL of this oil [47]. Because of its potential toxicity pennyroyal is not recommended for children and other sensitive groups [45].

In *Mentha pulegium* essential oil growing in Bosnia and Herzegovina, 34 essential oil components (98% components of the oil), have been identified by GC/MS and

**Figure 6.**
Mentha pulegium *L. (pennyroyal).*

GC/FID. Monoterpenoids were the most represented group of compounds (90.4%), with pulegone (54.4%), *p*-menthone (14.0%), piperitenone (12.8%), piperitone (3.7%), and isopulegone (2.5%) being dominant. Monoterpenes accounted for a total of 3.3%. The most prevalent monoterpenes were: limonene (1.2%), α-pinene (0.9%), and ß-pinene (0.6%). Of sesquiterpenes (1.7%), germacrene D was the most prevalent (1.1%). The non-terpenic compounds accounted for 2.5%, with 3-octanol being the most prevalent (2.3%) [48].

The antioxidant capacity of the essential oils of *Satureja subspicata* and *Mentha pulegium* were evaluated by the commonly used DPPH and FRAP assay. In DPPH test, in comparison to reference antioxidants ascorbic acid ($IC_{50}$ = 0.35 g/L) and hydroxy-anisole (BHA) ($IC_{50}$ = 0.37 g/L), essential oil of tested plants showed lower antioxidant potential for *Satureja subspicata* essential oil ($IC_{50}$ = 3.3 g/L) [40] and good antioxidant potential for *Mentha pulegium* essential oil ($IC_{50}$ = 94.3 μg/mL) [48].

The antioxidant potential of *S. subspicata* and *M. pulegium* essential oils, in concentration of 1 g/L tested by FRAP assay were for *S. subspicata* essential oil 73.89 (Eq $Fe^{2+}$μM), [40] and for *M. pulegium* essential oil 6.71 (Eq $Fe^{2+}$μM) [48]. Ascorbic acid and BHA had antioxidant potentials of 5568.43 and 5586.27 (Eq $Fe^{2+}$μM) respectively, for the same tested concentration [40, 48].

Low quantities of phenol compounds or monoterpenoids (such as carvacrol and thymol), which are good antioxidant compounds, may explain low antioxidant activity of *S. subspicata* essential oil [49]. As pulegone and menthone are known for their antioxidant properties, [50] the obtained results for *M. pulegium* essential oil can be explained by the high content of pulegone (54.4%) and the significant content of *p*-menthone (14.0%). Good antioxidant activity is also shown by 1,8-cineole, [51] which was also identified as one of the components of the tested *M. pulegium* oil (0.2%) [48].

The ability of *Satureja subspicata* and *Mentha pulegium* essential oils to inhibit enzymes acetylcholinesterase and butyrylcholinesterase (AChE and BChE) was tested by Ellman's method. The essential oils were tested at initial concentrations of

1 and 2 mg/mL. At an initial concentration of 0.1 mg/mL, eserine - the well-known cholinesterase (ChE) inhibitor - inhibited AChE with 95.9%, while BChE inhibited with 79.1%. In comparison to eserine, *S. subspicata* essential oil demonstrated good inhibitory activity of AChE at starting concentrations of 1 and 2 mg/mL (72.82% and 76.89%, respectively), and moderate inhibition of BChE (51,51% i 27,15%, respectively) [52]. *M. pulegium* essential oil at the same starting concentrations showed moderate inhibitory activity with inhibition of AChE (28.8 and 50.6%, respectively) and BChE (63.7 and 71.1%, respectively) [48].

These good results for *S. subspicata* essential oil may be due to presence of the well-known cholinesterase inhibitors α-pinene (10.2%) and β-caryophyllene (14%), among the main components. The moderate inhibitory activity of *M. pulegium* essential oil could be explained by the presence of moderate ChE inhibitor pulegone, as its primary component (54%). 1,8-cineole, a highly strong ACh inhibitor, is also present in *M. pulegium* oil in a smaller amount (0.2%), as well as α-pinene and ß-caryophyllene [51]. It is worth noting that our results reveal better inhibition of less specific BChE than AChE, which has to be further examined.

## 6. Conclusions

An excess of reactive species and damage caused by their action is called oxidative stress. In a state of oxidative stress, an excess of ROS and RNS may damage lipids, proteins, carbohydrates, and nucleic acids. Free radicals attack unsaturated fatty acids in biological membranes causing lipid peroxidation. Oxidative modification of proteins is present in diseases and changes associated with the aging process, such as atherosclerosis, tumors, neurodegenerative diseases, and the aging. In addition to fragmentation, oxidation of the amino acid residues increases carbonyl concentration, so the presence of carbonyl groups can be used as an indicator of protein oxidation.

Oxidation of lipid molecules is a major problem in the food industry, as it leads to changes in the organoleptic properties of food, a decrease in its nutritional value, as well as the formation of radical components that can endanger consumers' health. Polyunsaturated fatty acids oxidize much faster than monounsaturated or saturated ones. Lipid autooxidation is an autocatalytic reaction, which means that it progresses over time due to the formation of products that catalyze the reaction themselves.

The use of some synthetic antioxidants has negative effects on human health due to the promotion of carcinogenesis [10, 11], and there is a tendency to replace synthetic antioxidants, where possible, with non-toxic antioxidants of natural origin. More recently, essential oils have also been used as a substitute for synthetic antioxidants, in those food canning sectors where their use does not adversely affect product flavor [12].

Given that the oxidative stress is associated with the etiology and pathogenesis of many diseases, it is believed that eliminating the causes of oxidative stress may prevent or delay the occurrence of pathological changes and reduce the occurrence of diseases. Therefore, natural antioxidants – alone or in the form of extracts – may be useful in the treatment of such diseases. Thus the reason for the great interest in researching the antioxidant activity of aromatic, medicinal, and edible plants [13]. With the exception of some phenolic components, whose antimicrobial and antioxidant activity is well known, such data are not available for most other components of essential oils.

Inhibition of acetylcholinesterase prevents the hydrolysis of acetylcholine, thus prolonging its activity in the transmission of nerve impulses. This concept is applied in the treatment of diseases characterized by low ACh levels, such as Alzheimer's disease. Synthetic AChE inhibitors such as physostigmine, tacrine, and donepezil cause side effects such as hepatotoxicity and gastrointestinal disorders. This was an incentive for further research with the aim of finding new ChE inhibitors of natural origin, with greater efficiency and bioavailability, as well as with fewer side effects. Many phytochemicals are bioactive compounds, some of which show ChE inhibitory activity and represent a model for the development of new drugs, ChE inhibitors. As terpenes and terpenoids have shown relatively weak inhibitory capacity in studies published so far, it is necessary to develop analogues with an improved efficiency [21].

The obtained results show that the tested essential oils of *Satureja subspicata* and *Mentha pulegium* growing in Bosnia and Herzegovina contain compounds that, in addition to antioxidant activity, also show activity in terms of cholinesterase inhibition. Therefore, they may be important in the prevention and treatment of Alzheimer's disease and other neurodegenerative disorders, as well as in conditions of impaired homeostasis caused by oxidative stress, and also in food as protecting antioxidants [40, 48].

## Conflict of interest

The authors declare no conflict of interest.

## Author details

Mejra Bektašević[1]* and Olivera Politeo[2]

1 Biotechnical Faculty, University of Bihać, Bihać, Bosnia and Herzegovina

2 Faculty of Chemistry and Technology, University of Split, Split, Croatia

*Address all correspondence to: mejra_b@yahoo.com

## References


[1] Belitz HD, Grosch W. Food Chemistry. Berlin-Heidelberg: Springer Verlag; 1987. pp. 129-200

[2] Yen GC, Hsieh CL. Antioxidant activity of extracts from Du-Zhong (Eucommia ulmoides) toward various lipid peroxidation model in vitro. Journal of Agricultural and Food Chemistry. 1998;**46**:3952-3957. DOI: 10.1021/jf9800458

[3] Mayo JC, Tan DX, Sainz RM, Natarajan M, Lopez-Burillo S, Reiter RJ. Protection against oxidative protein damage induced by metal-catalyzed reaction or alkylperoxyl radicals: Comparative effects of melatonin and other antioxidants. Biochimica et Biophysica Acta. 2003;**1620**:139-150. DOI: 10.1016/s0304-4165(02)00527-5

[4] Wehr NB, Levine RL. In: Galluzzi L, Vitale I, Kepp O, Kroemer G, editors. Cell Senescence: Methods and Protocols. New York: Humana Press, Springer; 2013. pp. 265-281

[5] Harold William Schultz HW, editor. Symposium on Foods: Lipid and their Oxidation. Westport-Connecticut: The AVI Publishing Company Inc.; 1962. pp. 3-150

[6] McClements DJ, Decker EA. Lipids. In: Damodaran S, Parkin KL, Fennema OR, editors. Fennema's Food Chemistry. 4th ed. Boca Raton: CRC Press; 2008. pp. 155-217

[7] Shahidi F, Zhong Y. Lipid oxidation measurement methods. In: Shahidi F, editor. Bailey's Industrial Oil and Fat Products. John Wiley & Sons, Inc.; 2005. pp. 357-385. DOI: 10.1002/047167849X.bio050

[8] Martinez-Tome M, Jimenez AM, Ruggieri S, Frega N, Strabbioli R, Murcia MA. Antioxidant properties of Mediterranean spices compared with common food additives. Journal of Food Protection. 2001;**64**:1412-1419. DOI: 10.4315/0362-028x-64.9.1412

[9] Abdalla AE, Roozen JP. Effect of plant extracts on the oxidative stability of sunflower oil and emulsion. Food Chemistry. 1999;**64**:323-329. DOI: 10.1016/S0308-8146(98)00112-5

[10] Namiki M. Antioxidant Antimutagens in food. Critical Reviews in Food Science and Nutrition. 1990;**29**:273-300. DOI: 10.1080/10408399009527528

[11] Pokorny J. Natural antioxidants for food use. Trends in Food Science and Technology. 1991;**2**:223-227. DOI: 10.1016/0924-2244(91)90695-F

[12] Ruberto G, Baratta MT. Antioxidant activity of selected essential oil components in two lipid model systems. Food Chemistry. 2000;**69**:167-174. DOI: 10.1016/S0308-8146(99)00247-2

[13] Dorman HJ, Hiltunen R. Antioxidant and pro-oxidant in vitro evaluation of water-soluble food-related botanical extracts. Food Chemistry. 2011;**129**:1612-1618. DOI: 10.1016/j.foodchem.2011.06.017

[14] Rice-Evans CA, Miller NJ, Paganga G. Structure-antioxidant activity relationships of flavonoids and phenolic acids. Free Radical Biology & Medicine. 1996;**20**:933-956. DOI: 10.1016/0891-5849(95)02227-9

[15] Vladimir-Knežević S, Blažeković B, Kindl M, Vladić J, Lower-Nedza AD,

Brantner AH. Acetylcholinesterase inhibitory, antioxidant and phytochemical properties of selected medicinal plants of the Lamiaceae family. Molecules. 2014;**19**:767-782. DOI: 10.3390/molecules19010767

[16] Vladimir-Knežević S, Blažeković B, BivalŠtefan M, Babac M. Plant polyphenols as antioxidants influencing the human health. In: Rao V, editor. Phytochemicals as Nutraceuticals - Global Approaches to their Role in Nutrition and Health. Rijeka: InTech; 2012. pp. 155-190

[17] Kabera JN, Semanal E, Mussa AR, He X. Plant secondary metabolites: Biosynthesis, classification, function and pharmacological properties. The Journal of Pharmacy and Pharmacology. 2014;**2**:377-392

[18] Pinho BR, Ferreres F, Valentao P, Andrade PB. Nature as a source of metabolites with cholinesterase-inhibitory activity: An approach to Alzheimer's disease treatment. The Journal of Pharmacy and Pharmacology. 2013;**65**:1681-1700. DOI: 10.1111/jphp.12081

[19] Houghton PJ, Ren Y, Howes MJ. Acetylcholinesterase inhibitors from plants and fungi. Natural Product Reports. 2006;**23**:181-199. DOI: 10.1039/b508966m

[20] Roseiro LB, Rauter AP, MouratoSerralheiro ML. Polyphenols as acetylcholinesterase inhibitors: Structural specificity and impact on human disease. Nutrition and Aging. 2012;**1**:99-111. DOI: 10.3233/NUA-2012-0006

[21] Mukherjeea PK, Kumarb V, Malb M, Houghton PJ. Acetylcholinesterase inhibitors from plants. Phytomedicine. 2007;**14**:289-300. DOI: 10.1016/j.phymed.2007.02.002

[22] Alcolea-Palafox M, Posada-Moreno P, Ortuno-Soriano I, Pacheco-del-Cerro JL, Martinez-Rincon C, Rodriguez-Martinez D, et al. Research strategies developed for the treatment of Alzheimer's disease. Reversible and Pseudo-irreversible inhibitors of acetylcholinesterase: Structure-activity relationships and drug design. In: Atta-ur-Rahman M, Choudhary I, editors. Drug Design and Discovery in Alzheimer's Disease. Vol. 6. Sharjah: Bentham Science Publishers; 2014. pp. 426-477. DOI: 10.1016/c2014-0-04317-1

[23] Dhivya PS, Sobiya M, Selvamani P, Latha S. An approach to Alzheimer's disease treatment with cholinesterase inhibitory activity from various plant species. International Journal of PharmTech Research. 2014;**6**:1450-1468

[24] Burčul F, Blažević I, Radan M, Politeo O. Terpenes, Phenylpropanoids, Sulfur and other essential oil constituents as inhibitors of Cholinesterases. Current Medicinal Chemistry. 2020;**27**:4297-4343. DOI: 10.2174/0929867325666180330092607

[25] Politeo O, Bektašević M, Carev I, Jurin M, Roje M. Phytochemical composition, antioxidant potential and cholinesterase inhibition potential of extracts from *Menthapulegium* L. Chemistry & Biodiversity. 2018;**15**:e201800374. DOI: 10.1002/cbdv.201800374

[26] Jukić M, Politeo O, Maksimović M, Miloš M, Miloš M. In vitro acetylcholinesterase inhibitory properties of thymol, carvacrol and their derivatives thymoquinone and thymohydroquinone. Phytotherapy Research. 2007;**21**:259-261. DOI: 10.1002/ptr.2063

[27] Politeo O, Botica I, Bilušić T, Jukić M, Carev I, Burčul F, et al.

Chemical composition and evaluation of acetylcholinesterase inhibition and antioxidant activity of essential oil from Dalmatian endemic species Pinusnigra Arnold ssp. dalmatica (Vis.) Franco. Journal of Medicinal Plant Research. 2011;**5**:6590-6596. DOI: 10.5897/JMPR10.289

[28] Miyazawa M, Yamafuji C. Inhibition of acetylcholinesterase activity by bicyclic monoterpenoids. Journal of Agricultural and Food Chemistry. 2005;**53**:1765-1768. DOI: 10.1021/jf040019b

[29] Savelev SU, Okello EJ, Perry EK. Butyryl- and acetyl-cholinesterase inhibitory activities in essential oils of Salvia species and their constituents. Phytotherapy Research. 2004;**18**:315-324. DOI: 10.1002/ptr.1451

[30] Šilić Č. Endemičnebiljke. Sarajevo: Svjetlost; 1990

[31] Redžić S. Wild medicinal plants and their usage in traditional human therapy (southern Bosnia and Herzegovina, W. Balkan). Journal of Medicinal Plant Research. 2010;**4**:1003-1027. DOI: 10.5897/JMPR09.254

[32] Skočibušić M, Bezić N, Dunkić V. Phytochemical composition and antimicrobial activities of the essential oils from *Satureja subspicata* vis. Growing in Croatia. Food Chemistry. 2006;**96**:20-28. DOI: 10.1016/j.foodchem.2005.01.051

[33] Ćavar S, Maksimović M, Šolić ME, Jerković-Mujkić A, Bešta R. Chemical composition and antioxidant and antimicrobial activity of two Satureja essential oils. Food Chemistry. 2008;**111**:648-653. DOI: 10.1016/j.foodchem.2008.04.033

[34] Radonić A, Miloš M. Chemical composition and *in vitro* evaluation of antioxidant effect of free volatile compounds from *Saturejamontana* L. Free Radical Research. 2003;**37**:673-679. DOI: 10.1080/1071576031000105643

[35] Abad MJ, Bermejo P, Gonzales E, Iglesias I, Irurzun A, Carrasco L. Antiviral activity of Bolivian plant extracts. General Pharmacology. 1999;**32**:499-503. DOI: 10.1016/s03063623(98)00214-6

[36] Skočibušić M, Bezić N. Chemical composition and antidiarrhoeal activities of winter savory (*Saturejamontana*) essential oil. Pharmaceutical Biology. 2003;**41**:622-626

[37] Hajhashemi V, Sadraei H, Ghannadi AR, Mohseni M. Antispasmodic and anti- diarrhoeal effect of *Saturejahortensis* L. essential oil. Journal of Ethnopharmacology 2000;71:187-192. DOI: 10.1016/s0378-8741(99)00209-3

[38] Amanlou M, Dadkhah F, Salehnia A, Farsam H, Dehpour A. An anti- inflammatory and anti-nociceptive effects of hydro alcoholic extract of *Satureja khuzistanica* Jamzad extract. Journal of Pharmaceutical Sciences. 2005;**8**:102-106

[39] Ruberto G, Baratta MT. Antioxidant activity of selected essential oil components in two lipid model systems. Food Chemistry. 2000;**69**:167-174. DOI: 10.1016/S0308-8146(99)00247-2

[40] Bektašević M, Carev I, Roje M, Jurin M, Politeo O. Phytochemical CompositionAnd antioxidant activities of the essential oil and extracts of *Satureja subspicata* Vis. Growing in Bosnia and Herzegovina. Chemistry & Biodiversity. 2017;**14**:e1700239. DOI: 10.1002/cbdv.201700239

[41] Burnie G, Forrester S, Greig D, Guest S. Botanica. München: Neografia; 2003

[42] Burk DR, Senechal-Willisa LC, Lopez BG, Hogue S, Daskalova M. Suppression of lipopolysaccharide-induced inflammatory responses in RAW 264.7 murine macrophages by aqueous extract of Clinopodium vulgare L. (Lamiaceae). Journal of Ethnopharmacology. 2009;**126**:397-405. DOI: 10.1016/J.Jep.2009.09.026

[43] Brahmi F, Hauchard D, Guendouze N, Madani K, Kiendrebeogo M, Kamagaju L, et al. Phenolic composition, *in vitro* antioxidant effects and tyrosinase inhibitory activity of three Algerian *Mentha*species: *M. spicata*(L.), *M. pulegium*(L.) and *M. rotundifolia*(L.) Huds (Lamiaceae). Industrial Crops and Products. 2015;**74**:722-730. DOI: 10.1016/j.indcrop.2015.04.038

[44] Marzouk B, Ben Hadj Fredj M, Chraief I, Mastouri M, Boukef K, Marzouk Z. Chemical composition and antimicrobial activity of essential oils from Tunisian Menthapulegium L. Journal of the Science of Food and Agriculture. 2008;**6**:78-82

[45] Barceloux DG. Medical Toxicology of Natural Substances. Hoboken, New Jersey: John Wiley & Sons Inc. Publication; 2008. pp. 563-567. DOI: 10.1002/9780470330319

[46] Khosravi Zanjani MA, Mohammadi N, Zojaji M, Bakhoda H. Chemical composition of the essential oil of *Menthapulegium* L. and its antimicrobial activity on *Proteus mirabilis, Bacillus subtilis* and *Zygosaccharomycesrouxii*. Journal of Food Science and Technology. 2015;**2**:31-40

[47] Anderson IB, Mullen WH, Meeker JE, Khojasteh-Bakht SC, Oishi S, Nelson SD, et al. Pennyroyal toxicity: Measurement of toxic metabolite levels in two cases and review of the literature. Annals of Internal Medicine. 1996;**124**:726-734. DOI: 10.7326/0003-4819-124-8-199604150-00004

[48] Bektašević M, Politeo O, Carev I. Comparative study of chemical composition, cholinesterase inhibition and antioxidant potential of Menthapulegium L. essential oil. Chemistry & Biodiversity. 2021;**18**:e2000935. DOI: 10.1002/cbdv.202000935

[49] Bakkali F, Averbeck S, Averbeck D, Idaomar M. Biological effects of essential oils - a review. Food and Chemical Toxicology. 2008;**46**:446-475. DOI: 10.1016/j.fct.2007.09.106

[50] Teixeira B, Marquesa A, Ramos C, Batista I, Serrano C, Matos O, et al. European pennyroyal (*Menthapulegium*) from Portugal: Chemical composition of essential oil and antioxidant and antimicrobial properties of extracts and essential oil. Industrial Crops and Products. 2012;**36**:81-87. DOI: 10.1016/j.indcrop.2011.08.011

[51] Mata AT, Proenc C, Ferreira AR, Serralheiro ML, Nogueira JM, Araujo ME. Antioxidant and antiacetylcholinesterase activities of five plants used as Portuguese food spices. Food Chemistry. 2007;**103**:778-786. DOI: 10.1016/j.foodchem.2006.09.017

[52] Bektašević M. Chemical and Biological Profile of Selected Plant Species Used in Traditional Medicine of Bosnia and Herzegovina [Thesis]. Croatia: University of Split; 2018

Chapter 7

# Pharmaceutical and Therapeutic Potentials of Essential Oils

*Ishrat Nazir and Sajad Ahmad Gangoo*

## Abstract

It is a common perspective that medicinal plants have played and continue to perform an undeniably major role in the lives of people worldwide. Essential oils are the key constituents of medicinal herbs and their biological activities have been discovered since ancient times and are enormously utilised in multiple industries. The essential oils possess important biological properties like antibacterial, antioxidant, antiviral, insecticidal, etc. Because of these unique features they are more acceptable and are utilised in various fields throughout the world. In the cosmetics industry they play an important role in the development of perfumes while in the food industry they have been used as food preservatives. Essential oil components are interestingly utilised for pharmaceutical applications. The most investigated properties are anti-oxidant, anti-inflammatory, antimicrobial, wound-healing, anxiolytic activities etc. The current thrust area is evaluation for aromatherapy and anti-cancer, as it is noted that essential oils reported in plants may prevent, inhibit, or even reverse formation of cancerous cells. The aim of this chapter is to provide a concise and comprehensive overview on the therapeutic and pharmaceutical potential of essential oils in the current scenario.

**Keywords:** essential oil, therapeutic use, pharmaceutical potential

## 1. Introduction

The plants are the main source of food, clothing and shelter. Besides, different materials derived from plants are utilised in treatment against numerous ailments. Due to detrimental effects of synthetic medicines, the herb derived medicines are undergoing revival because of their safe application. Aromatic plants are the source of essential oils, which are volatile substances having essence and properties of the source plant.

The essential oils have been extracted from 60 families of plants from different parts of the world and around 3000 diverse essential oils have been recognised so far. Out of these around 300 are utilised monetarily in the seasoning and scents advertised [1]. The essential oils can be produced in all parts of a plant, mainly by leaves, flowers and stems (Peppermint, Lavender), fruits (Anise), bark (Cinnamon), seeds (Nutmeg). Plants store these components in the glandular cells or pockets which release them with aroma when squeezed or pressed [2]. Essential oil can be extracted by conventional methods: steam distillation, water and steam distillation. However, the cold or hot pressing, aqueous infusion, solvent extraction, effleurage or the other methods used for extraction of essential oils [3]. Bowles [4] reported that in some cases essential oil content may reach above 10% viz. Nutmeg (*Myristica fragrans*) and clove (*Syzygium aromaticum*) but in general the essential oil content rarely exceeds 1%. Essential oils possess physical properties as they are commonly hydrophobic in nature depicting slight solubility in water, although solubility in non- polar solvents varies, from highly soluble in waxes, alcohol and other weakly polar solvents. Further, the essential oils are commonly pale yellow or colourless but Chamomile (*Matricaria chamomilla*) essential oil is blue in colour. Moreover, they exist mainly in liquid state showing lower density than water except Sassafras, Cinnamon and Clove essential oil which are denser than water. [5, 6]. The principle chemical components are monoterpenes, sesquiterpenes, oxygenated derivatives, aromatic and aliphatic compounds. The complex mixtures of chemical compounds generally comprise of terpenoids, alcohols, ethers, asters, ketones and aldehydes in differential concentrations.

The pharmaceutical and therapeutic properties of plants are attributed to the essential oils and are related to their chemical composition [1]. All over the world researchers have established various pharmaceutical and therapeutic properties of the essential oils from time to time [7–9]. Extensive work has been carried out to utilise essential oils for the cure of multiple infectious diseases through pharmaceutical remedies. Scientific investigations have established that qualitatively 100% pure essential oil free from impurities have the potential to relieve chronic pain, elevate moods, recover defective cells and treat life threatening diseases, common in the world. The broad therapeutic prospective of the plant derived, essential oils have grab attention of the researchers all around to visualise their anti-cancer properties because of the fact that their mode of action is quite diverse than the classic cytotoxic chemotherapeutic agents [10]. Besides, one fascinating feature is their potential as medicines in aroma based therapies or as carriers for drug delivery. In the recent past the aim of essential oils have alternately shifted from culinary use to pharmaceutical and therapeutic use, yet in addition to their application in the fabrication of fragrances and beauty care products [11].

In the current scenario essential oils are gaining importance day by day, reason for this being they are mostly utilised in beverages, food industries, cosmetics and Fragrances industries for making valuable perfumes, beautifying agents, soaps, shampoos or cleaning gel. Also the significant contribution of essential oils is their utilisation in the Agro-food businesses for increasing the sensorial characters of food items [12]. The purpose of this chapter is to make an effort to bring the remedial and pharmaceutical significance of essential oils in light. For this purpose the recent research carried out throughout the world by various researchers has been included in this chapter.

## 2. Chemical composition of the essential oil

The chemical composition of essential oils varies plant to plant, the constituents of essential oils are generally volatile and non-volatile in nature therefore are widely categorised into volatile and non-volatile types. Further, the volatile fractions of aromatic oils are chemically constituted by the mono and sesquiterpene components and several oxygenated derivatives along with alcohols, aliphatic aldehydes, and esters, while as the non-volatile fractions are chemically constituted by the carotenoids, fatty acids, flavonoids and waxes [13].

The chemical composition of essential oils is determined by gas chromatography-mass spectrometry (GC-MS). This method is simple, efficient and gives fast results. Further, it is a broadly used analytical technique for the determination of essential oils constituents. A GC-MS provides a valid profile of the essential oil components and serves as the fingerprint of any particular batch of essential oil. The peculiar properties of the oils can be reflected from its chemical composition and GC-MS is a reliable technique to indicate the purity of essential oils in most cases [14]. The components of essential oils are delineated below.

## 2.1 Classes of essential oil compounds and their biological activities

### *2.1.1 Hydrocarbon*

The largest group of composites present in essential oils are hydrocarbons. The hydrocarbons are composed of carbon and hydrogen bits. The hydrocarbons which are found in essential oils are placed in a group called Terpenes (monoterpenes: C10, sesquiterpenes: C15, and diterpenes: C20). On the basis of physical composition the Terpenes may be ambrosial, alicyclic (monocyclic, bicyclic or tricyclic) or acyclic. The terpenes which are ingredients of essential oils are β-pinene, α-sabinene, myrcene, α-pinene, p-cymene, myrcene, α–phellandrene, pmenthane, thujane, fenchane, Limonene,, azulene, cadinene, sabineine and farnesene These composities have been associated with various remedial conditioning (**Table 1**).

### *2.1.2 Esters*

Esters are the chemical composites constituting an organic or inorganic acid with one hydroxy group replaced by an alkyl group. The Esters are generally found

| Compound | Chemical nature | Activity | References |
|---|---|---|---|
| Dodecane, phellandrene | Hydrocarbon | Antimicrobial activity | [15] |
| Bornanone | Terpene | Anti-inflammatory, antifungal, antimicrobial, anticancer | [15] |
| α- pinene, β- pinene, sabinene, myrecene, β- ocimene | Monoterpenes | Antimicrobial, antifungal, antioxidants | [16] |
| Eucalyptol, citronelal, eucamalol, linalool, α-terpineol | Alcohols | Antioxidant, insecticide, acaricide, herbeside | [17] |
| Bisabolane, elemane, germacrane, humulane, chamazulene | Sesquiterpenes | Antibacterial, antifungal, anti-inflammatory, antioxidant | [18] |
| Cinnanaldehyde, benzaldehyde, myrtenal | Aldehyde | Antifungal, circulatory, anti-inflammatory, cardiovascular | [19] |
| Geranyl acetate, bornyl acetate, linalyl acetate, eugenol acetate | Ester | Antispasmodic, antimicrobial | [20] |
| Pulegone, fenchone, thujone | Ketone | Antiviral, gastrointestinal, regenerating cells, analgesic | [21] |
| Carvacrol, tymol | Phenol | Antibacterial, strengthening immune system | [22] |

**Table 1.**
*Chemical nature of essential oil compounds and some biological activities.*

composites in a vast number of the essential oils and are known for their affable smell and give sweet smell to the essential oils. The common ester bearing essential oils include linalyl acetate, geraniol acetate, eugenol acetate and bornyl acetate. Esters are anti-inflammatory, spasmolytic, dreamy, and antifungal (**Table 1**).

#### *2.1.3 Alcohols*

Alcohol containing essential oils has a affable type of fragrance. The alcohol bearing essential oils are therapeutically most profitable essential oil components with no reported contraindications. Linalool, menthol, borneol, santalol, nerol, citronellol and geraniol are some important alcohols found in the essential oils. They are known to retain antimicrobial, antiseptic, tonifying, balancing and spasmolytic parcels (**Table 1**).

#### *2.1.4 Phenols*

These are aromatic alcohols which are chemically veritably reactive, slightly poisonous and induce irritation to the skin and the mucous membranes. They exist as crystals at room temperature. The important essential oils containing phenol s are thymol, eugenol, carvacrol and chavicol. The essential oils containing phenols as their constituents possess following characteristics, antimicrobial, rubefacient properties, stimulate the immune and nervous systems and may reduce cholesterol (**Table 1**).

#### *2.1.5 Ketones*

Ketones such as carvone, menthone, pulegone, fenchone, camphor, thujone and verbenone are some common examples of ketones found in essential oils. These groups of compounds are chemically stable and lack fragrance or flavour like the other group of compounds. Besides some remidial effects, Ketones have been reported to retain neurotoxic and abortifacient effects in some cases similar as camphor and thujone [23]. These ketone bearing essential oils have been reported to be mucolytic, cell regenerating; opiate, antiviral, analgesic and digestive in nature (**Table 1**).

#### *2.1.6 Aldehydes*

Aldehydes found in essential oils include citral (geranial and neral), myrtenal, cuminaldehyde, citronellal, cinnamaldehyde and benzaldehyde. Unlike ketones aldehydes retain sweet, pleasant fruity odours and are present in common culinary herbs such as cumin and cinnamon. This group of compounds are unstable and oxidise easily, besides numerous of the aldehydes have been reported to act as mucous membrane irritants and are skin sensitizers. As far as therapeutic use is concerned, aldehydes have been reported to work as antiviral, antimicrobial, tonic, vasodilators, hypotensive, calming, antipyretic and spasmolytic (**Table 1**).

## 3. Mechanism of action of bioactive components of essential oils

The mode of action of essential oils varies. The mode of action depends upon chemical composition and molecular structure of the components of essential oil.

### 3.1 Antibacterial action

An important feature of essential oils are their hydrophobicity, which allows them to partition into lipids of the cell membrane of bacteria disrupting the structure thus making it more permeable resulting in leakage of ions and cellular molecules which cases greater loss of cell contents leading to cell death for instance trans-cinnamaldehyde can inhibit the growth of E. coli and *Salmonella typhimurium*. It has been reported that essential oils containing primarily aldehydes and phenols for example cinnamaldehyde, citral, carvacrol, eugenol andthymol are characterised by maximum antibacterial activity followed by essential oils consisting of terpene- alcohols.

### 3.2 Antifungal action

Antifungal actions resemble in mode of action as those described for bacteria. In case of yeast it has been reported that potential of Hydrogen (pH) gradient across the cytoplasm membrane and blockage of energy production in the cells results in disruption of fungal membranes leading to death. Antifungal effects were caused by a combination of essential oils of clove and *rosmarinus officinalis* against *C. albicans*. Trans-anthole, a major component of Anise essential oil, demonstrated anti-fungal activity against the filamentous fungus, *Mucor mucedo*. The essential oil obtained from citrus containing active component limonene has been reported to inhibit the growth of *Aspergillus niger* by causing deleterious morphological alterations that is loss of cytoplasm fungal hyphae and budding of hyphal tip [24]. Also, tea tree essential oil containing components has been reported to alter permeability as well as membrane fluidity of *Candida albicans* [25].

### 3.3 Antiviral activity

The essential oil of saltolinia showed antiviral activity against HSV-1 and HSV-2 by preventing cell to cell virus spread in infected cells. The oil directly inactivated virus particles thus preventing adsorption of virion to host cells. Iso-borneol, a common monoterpene alcohol, showed dual virucidal activity against HSV-1, specifically inhibited glycosylation of viral polypeptides. The antiviral activity of the essential oil is principally due to direct virucidal effects (by denaturing viral structural proteins or glycoproteins). Proposed mechanisms suggest that essential oils intrude with the virus envelope by inhibiting specific processes in the viral replication cycle or by masking viral factors, which are necessary for adsorption or entry into host cells, therefore precluding cell-to-cell virus prolixity [26]. The essential oils attained from oregano and clove have been reported to show remarkable antiviral exertion against a number of non-enveloped DNA and RNA viruses including adenovirus type-3, coxsackievirus B-1 and polio virus. Several constituents of essential oils like monoterpenes, sesquiterpenes and triterpenes have been reported to show strong antiviral activity against rhinovirus and herpes virus. The essential oil components of pogostemoncablin have been found active against H2N2 influenza-A virus [27].

### 3.4 Anticancer activity

The broad therapeutic prospective has gained a lot of attention throughout the world in recent times for their implicity capacity in relation to combating cancer. According to Wu et al. [28] diallyl sulphide, diallyldisulfide composites actuated in the host cells (rats) the enzymes which play an important part in the detoxification process of hepatic phase-1 (decomposition of chemical bonds that link carcinogenic toxins to each other) and phase-2 (bonds to toxins released detoxifying enzymes similar as glutathione S- transferase). Further myristicin an allyl benzene composites found in the essential oil of nutmeg activates glutathione S- transferase in mice cells which minimise carcinogenesis induced by benzo a pyrene in the lungs of mice. Moreover it has been recently concluded that myristicin persuade apoptosis in neuroblastoma (SK-N-SH) in humans [18]. Geraniol have been reported to decrease the resistance of: cancer cells (TC 118) to 5-fluorouracil an anticancer agent. Further, geraniol enhances the inhibitory effect of tumour growth 5-fluorouracil. Moreover the essential oil of balsam fir which contains alpha-humulene depicting high anticancer property in several cell lines and low toxicity to healthy cells [29]. In addition to this limonene an active component of citrus essential oil has been reported to show anticancer activity at the level of stomach cancer and liver cancer [30]. Chamomile essential oil containing an active component alpha-bisabolol sesquiterpene alcohol has been reported to show antigliomale activity [31].

## 4. Therapeutic properties of some essential oils

### 4.1 Chamomile essential oil (*Matricariachamomilla*)

It has been reported that Bisabolol and chamazulene are active compounds found in chamomile essential oil. The dry flowers of Chamomile have numerous properties such as anti-inflammatory, antioxidant and also possess some mild astringent properties [32–35].

### 4.2 Anise essential oil (*Pimpinellaanisum*)

The Anetholeis is the main active compound found in ansine essential oil. Therapeutic Properties of Anise include a cure for sleeplessness, an appetite stimulant and diuretic. In ancient times the Anise has been reported to show the carminative property (reducing flatulence) [36–38].

### 4.3 Nutmeg essential oil (*Myristicafragrans*)

The Main active compounds found in nutmeg essential oil are Sabinene, 4-terpineol and myristicin. The essential oil of Nutmeg has been found effective against a number of microbial agents and pests. Also, it is used as an important ingredient to cough syrups, while in some instances it acts as general tonic for brain activity and normal functioning of circulatory system [39].

### 4.4 Cedar essential oil (*Cedruslibani*)

Cetin et al. [40] reported that the principle active component of cedar essential oil is Limonene. The essential oil of cedar has been found to perform Antifungal and Larvicidal activity. Also, the oil is good for regeneration of blood cells and enhances the healing property [41–43].

### 4.5 Dill essential oil (*Anethumgraveolens*)

The Main active compound found in dill essential is Carvone and the well-established therapeutic use of essential oil reported is the Antispasmodic in gastrointestinal disorders. Moreover, it reduces the fluidity of bronchial secretions in the lungs and thus prevents various lung infections [21, 44].

### 4.6 Garlic essential oil (*Allium sativum*)

It has been reported that Diallylle disulphide is the main active compound found in garlic essential oil. Garlic essential oil Protects and maintains the cardiovascular system, reducing blood pressure. Also, the extracted essential oil has been reported to control the fungal infection, pest infestation and parasitic growth. Moreover, many studies have found an increase in garlic intake reduces the cancers of the upper digestive tract [45].

### 4.7 Clove essential oil (*Syzygiumaromaticus*)

Eugenol and eugenyl acetate are the main active compounds which constitute clove essential oil. The commonly known therapeutic property of essential oil reported is effectiveness against the tooth ache and as an analgesic for alveolar osteitis. Also, the studies have proven that essential oil obtained from clove is effective against various microbial and fungal infections [46, 47].

### 4.8 Cinnamon essential oil (*Cinnamomum cassia*)

Essential oil of Cinnamon is mainly constituted by the chemical Cinnamaldehyde. The cinnamon essential oil has been reported to perform enormous functions related to health. The studies have established that it lowers the plasma glucose in the diabetic patients. Also, it has been reported to lower the level of total cholesterol and triglycerides in the blood, thus preventing cardiovascular diseases [48, 49].

### 4.9 Sweet orange essential oil (*Citrus sinensis*)

Limonene is Main active compound found in sweet orange essential oil. It possesses. Antiseptic property in some cases but the commonly reported property of essential oil is used as an excellent flavouring ingredient in the food industry [50, 51].

### 4.10 Eucalyptus essential oil (*Eucalyptus globulus*)

A number of compounds has been reported in the eucalyptus essential oil but 1,8-cineole is the major constituent present in the essential oil. A number of studies have concluded that it can be used for treating cough, common cold and to mildly relieve muscular pain. Also, the essential oil is used as an insect repellent and biopesticide in many countries [52, 53].

### 4.11 Peppermint essential oil (*Menthapiperita*)

The major portion of essential oil contains menthol and menthone compounds, which govern the properties like treatment for irritable bowel syndrome. Also, used topically for muscle pain, nerve pain and relief from itching in many cases. Moreover, it has been found to minimise the mucosal irritation in the digestive tract and reduce the heartburn [51, 54, 55].

### 4.12 Lavender essential oil (*Lavandulaofficinalis*)

Linalool and linalyl acetate are main active compounds found in lavender essential oil. The main properties of lavender essential oil includes sedative action, pan relaxing, analgesic in many cases and effective in alleviating in anxiety and sleep disturbances [56–59].

### 4.13 Tea tree essential oil (*Melaleucaalternifolia*)

The main active compound reported so far in the essential oil has been only Terpinène-1-ol-4. The tea tree essential oil has been used to treat coughs and colds widely. In addition, the oil is used to treat sore throats and numerous skin ailments [60–62].

### 4.14 Lemon essential oil (*Citrus limonum*)

Limonene is the main constituent compound found in lemon essential oil. The therapeutic Properties include enhancement of natural immunity in the human body, regulation of metabolism and a reliable nerve tonic. Besides, it has been concluded through many studies that essential oil acts as antiviral and antimicrobial [63–66].

### 4.15 Yarrow (*Achillea millefolium*)

The important active constituents of yarrow essential oil are Sabinene and terpineol manifesting. A number of studies have reported that the essential oil of yarrow acts as an anti-inflammatory and analgesic. Moreover, it has been found to cure many lung diseases and act as an important antiseptic agent [67, 68].

### 4.16 Geranium (*Pelargonium graveolens*)

The geranium essential oil consists of citronellol, geraniol, linalool and citronellylformate. The essential oil has been found to depict astringent and antiseptic properties. Also, in minor instances anti-inflammatory and antioxidant property has been observed [69].

### 4.17 Thyme (*Thymus vulgaris*)

Chromatographic analysis has revealed that the main active compound in essential oil is thymol followed by carvacrol, linalool etc. Thymol shows antiseptic properties and is an active ingredient of commercially prepared mouthwashes and toothpastes.

## 5. Conclusion

This chapter comprehensively summarises the therapeutic and pharmaceutical potential of essential oils. The essential oils possess important biological activities which lead to their application in diverse fields. The characteristic properties such as antiviral, anti-bacterial, anti-fungal, anti-inflammatory, ant carcinogenic etc. are utilised in various industries to prepare beneficial products which have great impact on human life. The active compounds present in essential oils are thoroughly studied now a day for replacement to unsafe medications. In pharmaceutical industries the use of essential for making perfumes and other pharmaceutical products are gaining popularity. Therefore, the essential oils are receiving attention from all the corners because of their tremendous features. Thus essential oils and their constituents can arguably be studied in the future for meticulously more scientific investigations and probable applications as important components in future medical field and pharmaceutical industries.

## Author details

Ishrat Nazir* and Sajad Ahmad Gangoo
Faculty of Forestry, SKUAST-Kashmir, Benhama, Jammu and Kashmir, India

*Address all correspondence to: ishratnazir12345@gmail.com

## References


[1] Raut JS, Karuppayil SM. A status review on the medicinal properties of essential oils. Industrial Crops and Products. 2014;**62**:250-264

[2] Abu-shanab SB, Adwan GM, Abu-Safiya D, Jarrar N, Adwan K. Antibacterial activities of some plant extracts utilized in popular medicine in Palestine. Turkish Journal of Biology. 2005;**28**:99-102

[3] Surburg H, Panten J. Common Fragrance and FlavorMaterials. Preparation, Properties and Uses. 5th ed. Weinheim: WILEY-VCH; 2006

[4] Bowles EJ. The Chemistry of Aromatherapeutic Oils. 3rd ed. New York: Edition Griffin Press; 2003

[5] Gupta V, Mittal P, Bansal P, Khokra SL, Kaushik D. Pharmacology potential of *matricaria recutita*—A review. International Journal of Pharmaceutical Sciences and Drug Research. 2010;**2**:12-16

[6] Martín A, Varona S, Navarrete A, Cocero MJ. Encapsulation and co-precipitation processes with supercritical fluids: Applications with essential oils. The Open Chemical Engineering Journal. 2010;**4**:31-41

[7] Buchbauer G, Jäger W, Jirovetz L, Limberger J, Dietrich H. Therapeutic properties of essential oils and fragrances. In: Teranishi R, Buttery RG, Sugisama H, editors. Bioactive Volatile Compounds from Plants. AC Symposium Series 525. Washington DC: American Chemical Society; 1993b

[8] Federspil P, Wulkow R, Zimmermann T. Effects of standardized Myrtol in therapy of acute sinusitis-results of a double-blind, randomized multicenter study compared with placebo. 705 Laryngorhinootologie. 1997;**76**:23-27

[9] Rajesh D, Howard P. Perillyl alcohol mediated radio-sensitization via augmentation of the Fast pathway in prostate cancer cells. Prostate. 2003;**57**:14-23

[10] Rajesh D, Stenzel R, Howard S. Perillyl alcohol as a radio-/chemosensitizer in malignant glioma. Journal of Biological Chemistry. 2003;**278**:35968-35935

[11] Naeem A, Abbas T, Ali TM, Hasnain A. Essential oils: Brief background and uses. Annals of Short Reports. 2018;**1**(1):1006

[12] Muhammad I, Muhammad AS, Saqib A, Amjad H. Biological Importance of Essential Oils. In: El-Shemy HA, editor. Essential Oils - Oils of Nature. IntechOpen; 17th December 2019. pp. 1-15. DOI: 10.5772/intechopen.87198

[13] Hussain A, Anwar F, Hussain SST, Przybylski R. Chemical composition, antioxidant and antimicrobial activities of basil (Ocimum basilicum) essential oils depends on seasonal variations. Food Chemistry. 2008;**108**:986-995

[14] Daferera DJ, Ziogas BN, Polissiou MG. GC-MS analysis of essential oils from some Greek aromatic plants and their fungitoxicity on Penicillium digitatum. Journal of Agricultural and Food Chemistry. 2000;**48**:2576-2581

[15] Padma M, Ganesan S, Jayaseelan T, Azhagumadhavan S, Sasikali P, Senthilkumar S, et al. Phythochemical screening and GC-MS analysis of bioactive compounds present in ethanolic

leaves extract of *Silybum marianum*. L. Journal of Drug Delivery and Therapeutics. 2019;**9**(1):85-89

[16] Aziz EE, Badawy ME, Zheljazkov DV, Nicola MS, Fouad H. Yield and chemical composition of essential oil of *Achillea milifollium* L. as affected by harvest time. Egypt Journal of Chemicals. 2019;**62**(3):933-940

[17] Almas I, Innocent E, Machumi F, Kisinza W. Chemical composition of essential oils from *Eucalyptus globulus* and *Eucalyptus maculate* grown in Tanzania. Scientific African. 2021;**12**:75-85

[18] Dhifi W, Bellili S, Jazi S, Bahloul N, Mnif W. Essential oils' chemical characterization and investigation of some biological activities: A critical review. Medicine. 2016;**3**(25):2-16

[19] Hamid AA, Aiyellagbe OO, Usman LA. Essential oils: Its medicinal and pharmacological uses. International Journal of Current Research. 2011;**3**(2):86-98

[20] Djilani A, Dicko A. the therapeutic benefits of essential oils, nutrition, well-being and health. In: Bouayed J, Bohn T, editors. IntechOpen; 2012. pp. 155-178. DOI: 10.5772/25344

[21] Edris AE. Pharmaceutical and therapeutic potentials of essential oils and their individual volatile constituents. Phytotherapy Research. 2007;**21**:308-323

[22] Ultee A, Bennik M, Moezelaar R. The phenolic hydroxl group of carvacrol is essential for action against the food born pathogen bcillus ceracus. Applied Environmental Microbiology. 2002;**68**:1561-1568

[23] Gali-Muhtasib H, Hilan C, Khater C. Traditional uses of *Salvia libanotica* (East Mediterranean sage) and the effects of its essential oils. Journal of Ethnopharmacology. 2000;**71**:513-520

[24] Gogoi P, Baruah P, Nath SC. Effects of *Citrus sinensis* (L.)Osbeckepicarp essential oil on growth and morphogenesis of *Aspergillusniger* (L.) Van Tieghem. Microbiological Research. 2008;**163**:337-344

[25] Hammer A, Carson F, Riley V, Nielsen B. A review of the toxicity of Melaleucaalternifolia (tea tree) oil. Food and Chemical Toxicology. 2006;**44**(5):616-625

[26] Saddi M, Sanna A, Cottiglia F, Chisu L, Casu L, Bonsignore L, et al. Antiherpevirus activity of Artemisia arborescens essential oil and inhibition of lateral diffusion in Vero cells. Annals of Clinical Microbiology and Antimicrobials. 2007;**6**:1-10

[27] Reichling J, Schnitzler P, Suschke U, Saller R. Essential oils of aromatic plants with antibacterial, antifungal, antiviral, and cytotoxic properties—An overview. Forschende Komplementärmedizin. 2009;**16**:79-90

[28] Wu CC, Sheen LY, Chen HW, Kuo WW, Tsai SJ, Lii CK. Differential effects of garlic oil and its three major organosulfur components on the hepatic detoxification system in rats. Journal of Agricultural and Food Chemistry. 2002;**50**:378-383

[29] Carnesecchi S, Langley K, Exinger F, Gossé F, Raul F. Geraniol, a component of plant essential oils, sensitizes human colonie cancer cells to 5-fluorouracil treatment. Journal of Pharmacology and Experimental Therapeutics. 2002;**301**:625-630

[30] Legault J, Dahl W, Debiton E, Pichette A, Madelmont JC. Antitumor activity of balsam fir oil: Production

of reactive oxygen species induced by Humulene as possible mechanism of action. Planta Medica. 2003;**69**:402-407

[31] Uedo N, Tatsuta M, Lishi H, Baba M, Sakai N, Yano H, et al. Inhibition by D limonene of gastric carcinogenesis induced by N methyl N' nitro N-nitrosoguanidine in wistar rats. Cancer Letters. 1999;**137**:131-136

[32] Alves AMH, Gonçalves JCR, Cruz JS, Araújo DAM. Evaluation of the sesquiterpene (−)-α-bisabolol as a novel peripheral nervous blocker. Neuroscience Letters. 2010;**472**:11-15

[33] Bnouham M. Medicinal plants with potential galactagogue activity used in the moroccan pharmacopoeia. Journal of Complementary and Integrative Medicine. 2010;7(1):52

[34] McKay DL, Blumberg JB. A Review of the bioactivity and potential health benefits of chamomile tea (Matricariarecutita L.). Phytotherapy Research. 2006;**20**:519-530

[35] Tolouee M, Alinezhad S, Saberi R, Eslamifar A, Zad SJ, Jaimand K, et al. Effect of Matricariachamomilla L. flower essential oil on the growth and ultrastructure of Aspergillus niger van Tieghem. International Journal of Food Microbiology. 2010;**139**:127-133

[36] Jaiswal P, Kumar P, Singh VK, Singh DK. Biological effects of myristicafragrans. Annual Review of Biomedical Sciences. 2009;**11**:21-29

[37] Nerio LS, Olivero-Verbel J, Stashenko EE. Repellent activity of essential oils from seven aromatic plants grown in Colombia against *Sitophiluszeamais* Motschulsky (*Coleoptera*). Journal of Stored Products Research. 2009;**45**:212-214

[38] Tabanca N, Demirci B, Ozek T, Kirimer N, Baser KHC, Bedir E, et al. Gas chromatographic–mass spectrometric analysis of essential oils from Pimpinella species gathered from Central and Northern Turkey. Journal of Chromatography A. 2006;**1117**:194-205

[39] Tomaino A, Cimino F, Zimbalatti V, Venuti V, Sulfaro V, De Pasquale A, et al. Influence of heating on antioxidant activity and the chemical composition of some spice essential oils. Food Chemistry. 2005;**89**:549-554

[40] Cetin H, Kurt Y, Isik K, Yanikoglu A. Larvicidal effect of (*Cedruslibani)* seed oils on mosquito Culexpipiens. Pharmaceutical Biology. 2009;**47**: 665-668

[41] Dharmagadda VSS, Naik SN, Mittal PK, Vasudevan P. Larvicidal activity of *Tagetespatula* essential oil against three mosquito species. Bioresource Technology. 2005;**96**: 1235-1240

[42] Kizil M, Kizil G, Yavuz M, Aytekin C. Antimicrobial activity of resins obtained from the roots and stems of *Cedruslibani* and *Abies Cilicia*. Applied Biochemistry and Microbiology. 2002;**38**(144):146

[43] Loizzo MR, Saab A, Tundis R, Statti GA, Lampronti IH, Menichini F, et al. Phytochemical analysis and in vitro evaluation of the biological activity against herpes simplex virus type 1 (HSV-1) of *Cedruslibani* A. Rich. Phytomedicine. 2008;**15**:79-83

[44] Bakkali F, Averbeck S, Averbeck D, Idaomar M. Biological effects of essential oils. Food and Chemical Toxicology. 2008;**46**:446-475

[45] Klevenhusen F, Zeitz JO, Duval S, Kreuzer M, Soliva CR. Garlic oil and its

principal component diallyldisulfide fail to mitigate methane, but improve digestibility in sheep. Animal Feed Science and Technology. 2011;**167**:356-363

[46] Koba K, Nenonene AY, Raynaud C, Chaumont JP, Sanda K. Antibacterial activities of the buds essential oil of *Syzygiumaromaticum* (L.) 42 Merr.& Perry from Togo. Journal of Biologically Active Products from Nature. 2011;**1**:42-51

[47] Machado M, Dinis AM, Salgueiro L, Custódio JBA, Cavaleiro C, Sousa MC. Anti-giardia activity of *Syzygiumaromaticum* essential oil and eugenol: Effects on growth, viability, adherence and ultrastructure. Experimental Parasitology. 2011;**127**:732-739

[48] Geng S, Cui Z, Huang X, Chen Y, Xu D, Xiong P. Variations in essential oil yield and composition during *Cinnamomum cassia* bark growth. Industrial Crops and Products. 2011;**33**:248-252

[49] Unlu M, Ergene E, Unlu GV, Zeytinoglu HS, Vural N. Composition antimicrobial activity and in vitro cytotoxicity of essential oil from *Cinnamomumzeylanicum* Blume (Lauraceae). Food and Chemical Toxicology. 2010;**48**:3274-3280

[50] Ezeonu FC, Chidume GI, Udedi SC. Insecticidal properties of volatile extracts of orange peels. Bioresource Technology. 2001;**76**:273-274

[51] Singh P, Shukla R, Prakash B, Kumar A, Singh S, Kumar P, et al. Chemicalprofile, antifungal, antiaflatoxigenic and antioxidant activity of *Citrus maxima* Burm. and *Citrussinensis* (L.) Osbeck essential oils and their cyclic monoterpene, dl-limonene. Food and Chemical Toxicology. 2010;**48**:1734-1740

[52] Ben-Arye E, Dudai N, Eini A, Torem M, Schiff E, Rakover Y. Treatment of upper respiratory tract infections in primary care: A randomized study using aromatic herbs. Evidence Based Complementary and Alternative Medicine. 2011;**690346**:7

[53] Gende L, Maggi M, Van Baren C, Leo D, Lira A, Bandoni A, et al. Antimicrobial and miticide activities of *Eucalyptus globulus* essential oils obtained from different Argentine regions. Spanish Journal of Agricultural Research. 2010;**8**:642-650

[54] Kumar P, Mishra S, Malik A, Satya S. Insecticidal properties of *Mentha* species. Industrial Crops and Products. 2011;**34**:802-817

[55] Sabzghabaee AM, Nili F, Ghannadi A, Eizadi-Mood N, Maryam AM. Role of menthol in treatment of candidal napkin dermatitis. World Journal of Pediatrics. 2011;**7**:167-170

[56] Kloucek P, Smid J, Frankova A, Kokoska L, Valterova I, Pavela R. Fast screening method for assessment of antimicrobial activity of essential oils in the vapour phase. Food Research International. 2012;**47**:162-165

[57] Pohlit AM, Lopes NP, Gama RA, Tadei WP, Neto VFD. Patent literature on mosquito repellent inventions which contain plant essential oils. Planta Medica. 2011;**77**:598-617

[58] Woronuk G, Demissie Z, Rheault M, Mahmoud S. Biosynthesis and therapeutic properties of *Lavandula* essential oil constituents. Planta Medica. 2011;**77**:7-15

[59] Zuzarte M, Gonçalves MJ, Cavaleiro C, Canhoto J, Vale-Silva L, Silva MJ, et al. Chemical composition

and antifungal activityof the essential oils of *Lavandula viridis* L'Hér. Journal of Medical Microbiology. 2011;**60**:612-618

[60] Garozzo A, Timpanaro R, Bisignano B, Furneri PM, Bisignano G, Castro A. *In vitro* antiviral activity of *M elaleuca alternifolia* essential oil. Letters in Applied Microbiology. 2009;**49**:806-808

[61] Lobo R, Prabhu K, Shirwaikar A, Shirwaikar A, Ballal M. Formulation and evaluation of antiseptic activity of the herbal cream containing *Curcuma longa* and tea tree oil. Journal of Biologically Active Products from Nature. 2011;**1**:27

[62] Mickienė R, Bakutis B, Baliukonienė V. Antimicrobial activity of two essential oils. Annals of Agricultural and Environmental Medicine. 2011;**18**: 139-144

[63] Koul O, Walia S, Dhaliwal GS. Essential oils as green pesticides: Potential and constraints. Biopesticides International. 2008;**4**:63-84

[64] Pavela R. Insecticidal properties of several essential oils on the house fly (*Muscadomestica* L.). Phytotherapy Research. 2008;**22**:274-278

[65] Pavela R. Insecticidal activity of some essential oils against larvae of *Spodopteralittoralis*. Fitoterapia. 2005;**76**:691-696

[66] Ponce AG, del Valle CE, Roura SI. Natural essential oils as reducing agents of peroxidase activity in leafy vegetables. LWT-Food Science and Technology. 2004;**37**:199-204

[67] Candan F, Unlu M, Tepe B, Daferera D, Polissiou M, Sokmen A, et al. Antioxidant and antimicrobial activity of the essential oil and methanol extracts of *Achillea millefolium* subsp. *millefolium* Afan. (Asteraceae). Journal of Ethnopharmacy. 2003;**87**:215-220

[68] Yaseen M, Ahmad M, Wani TA, Ahmad M, Gani BA, Qureshi R. Phytochemical screening and antioxidant activity of extracts of the leaf and stem of *Achillea millefolium*. International Journal of Science and Research. 2017;**2**:55-59

[69] Boukhris M, Bouaziz M, Feki J, Feki A, Sayadi S. Hypoglycemic and antioxidant effects of leaf essential oil of *Pelargonium graveolens* L. Hert. In alloxan induced diabetic rats. Lipids Health and Disease. 2012;**11**:81

Chapter 8

# Application of Essential Oils in the Treatment of Inflammatory Bowel Disease

*Najmeh Oliyaei, Nader Tanideh and Seyedeh Zahra Nasirifar*

## Abstract

Essential oils (EOs) are natural compounds obtained from algae and different parts of plants. EOs are volatile secondary metabolites and are classified into major groups, including terpenes/terpenoids and aromatic/aliphatic compounds. There are numerous studies about the biological activities of EOs, demonstrating their abilities for the prevention and treatment of diseases. Their biological activities are mainly related to their constituents, such as α-pinene, thymol, 1, 8-cineole, carvacrol, etc. Thus, the use of EOs as pharmaceutical agents for curing several diseases has gained much attraction in recent years. Moreover, inflammatory bowel disease (IBD) is a type of disease that causes chronic inflammation in the intestine. Ulcerative colitis (UC) and Crohn's disease (CD) are two main forms of IBD. Some studies have reported the efficacy of EOs in treating IBD, in particular, UC. This chapter will focus on the biomedical application of EOs in the treatment of IBD.

**Keywords:** bioactive compounds, essential oils, biological properties, inflammatory bowel disease, ulcerative colitis

## 1. Introduction

Inflammation is a physiological response against various infection agents, toxins, and injury which are related to several disorders [1]. Inflammatory bowel disease (IBD) and irritable bowel syndrome (IBS) are two common disorders with similar signs of abdominal pain but different pathophysiology and therapeutic methods. IBD causes inflammation of the intestines and is a term for a broad spectrum of diseases and crohn's disease (CD) and ulcerative colitis (UC) are the most common. IBD treatment usually needs a lifetime of medical care, while IBS affects the large intestine without inflammation [2, 3]. Moreover, IBD affects 6.8 million people worldwide, in particular in North America and United Kingdom [4]. The genetic predisposition, gut microbiota, environmental risk factors, and dysfunction of the immune system can be related to IBD. The IBD is characterized by intestinal epithelial injury (excess mucus production), inflammation expansion, and failed control of the inflammatory response [5]. Furthermore, it has been established that the gut microbiota has an important effect on IBD pathology as their close connection to the host immune system [6]. The usual approach and therapeutic management strategies for IBD are drug treatment by monoclonal antibodies, immunomodulators, corticosteroids, and aminosalicylates which may have some side effects [4]. Therefore, the development of a safe and effective strategy is required for the treatment of IBD patients [7]. The novel treatment of gastrointestinal diseases is utilizing natural bioactive compounds such as crude extracts, essential oils (EOs), and pure isolated compounds from medicinal plants with improving immune function attributes [8]. Thus, natural bioactive compounds with anti-inflammatory effects such as herbal medicine extracts [9, 10] and EOs have attracted much attention to develop new types of anti-inflammatory agents [11]. There are several investigations confirmed the anti-inflammatory effect of the herbal medicine extracts or EOs in different UC models animal models [12–15]. It has been established that EOs can improve the balance of gastrointestinal immunity by their anti-inflammatory activity and downregulation of pro-inflammatory products [16].

EOs are highly volatile secondary metabolites and are known as aromatic substances produced by specific plants [17] or algae species [18]. EOs have considerable potential to be used as a part of pharmaceuticals, nutraceuticals, and functional foods because of their broad range of biological activities [17]. EOs are involved in monoterpenes, sesquiterpenes, and oxygenated derivatives of these and possess synergistic effects in combination together. The usual extraction methods of EOs are steam distillation, hydrodistillation, or solvent extraction. However, there are numerous factors are known to influence the properties of EOs including species and genetic, climate, and geographic origin which caused differences in chemical structures [19]. Moreover, the different chemical structures of EOs exhibit different biological properties [20]. A growing interest in using natural bioactive compounds as medicine or food preservation results in increasing interest in EOs applications. They are characterized by their potential health benefits including antibacterial, antifungal, antiviral, insecticidal, and antioxidant activities. These attributes are related to single or groups of compounds, which play an important role in the defense mechanisms of plants against abiotic stress [21].

This chapter presents an overview of EOs potential effect in promoting health, in particular, IBD.

## 2. Inflammatory bowel disease (IBD)

IBD is a group of diseases that caused diarrhea, abdominal pain or discomfort, and even bloody stool. These inflammatory intestinal diseases are involving the ileum, rectum, and colon. The two main forms of IBD are UC and CD with different clinical, pathogenic, and biomolecular properties. Some investigations reported that IBD is a heterogeneous medical condition distinguished by inflammation of the gastrointestinal tract due to the unusual response of aggressive types of T-cells to luminal microbiota in genetically susceptible patients [3]. Several mediators involved in inflammation and immune responses are represented to impact on IBD, including pro-inflammatory cytokines including tumor necrosis factor (TNF), Interferons (IFN-γ), interleukins (IL-6, IL-12, IL-21, IL-23, IL-17) and anti-inflammatory cytokines (IL-10, TGFβ, IL-35, etc.). CD is usually characterized by an increased secretion of IL-12, IL-23, IFN-γ, and IL-17 by Th1 and Th17 cells while Th2 and Th9 cells are considered in UC by secretion of IL-13, IL-5, and IL-9. Several studies were investigated about cytokines effects in initiating, mediating, perpetuating, and controlling intestinal inflammation and tissue injury because they are the crucial parameters in the pathogenesis of IBD and may have potential therapeutic targets [5]. In addition, NLRP1, NLRP3, NLRC4, absent-in-melanoma 2, and pyrin (types of pattern-recognition receptors (PRRs)), construct inflammasomes. NLRP3, one of the NOD-like receptor family member, has been investigated more than other inflammasomes intimately pertinent to IBD. The principal clinical expressions of most patients with IBD consist of uncommon levels of the NLRP3 inflammasome and pro-inflammatory cytokines [22]. Moreover, oxidative stress has been shown to participate in major mechanisms of some disorders such as IBD. Uncontrolled and persistent oxidative and nitrosative stress with overproduction have an important effect on chronic disorders such as IBD as can be seen in **Figure 1** [24]. The principal ROS consist of superoxide anion ($O_2$), nitric oxide (NO), hydroxyl radical ($^-OH$), hydroperoxyl radical ($O_2H$), hydrogen peroxide ($H_2O_2$), and singlet oxygen ($O_2$) [25]. Antioxidant equilibrium can eliminate the harmful effects of ROSs and RONs. They classify intracellular and plasma antioxidant mechanisms [23].

## 2.1 Risk factors

There are several risk factors of IBD, including lifestyle, age, genetic and immune response. Diet, in particular, meats and oily foods, and exposure to different contaminations and antibiotics may change the gut microbiota resulting in IBD. While consumption of vegetables, fruits, and fish can reduce the risk of IBD. Indeed, various food products may change gut permeability and cause dysfunctional intestinal mucosa [26]. Patients diagnosed with IBD also have the lower health-related quality of life and risk factor for colon cancer [27]. Meanwhile, it has been established that functional foods rich in grape seed oil [28] extra virgin olive oil, canola oil, and rice bran oil [29] or herbal medicine extract such as *Pistacia atlantica* [27] are preferable means for overcoming the limitations of the current drug treatments of UC. In addition, various intestinal immune cells are responsible against foreign antigens

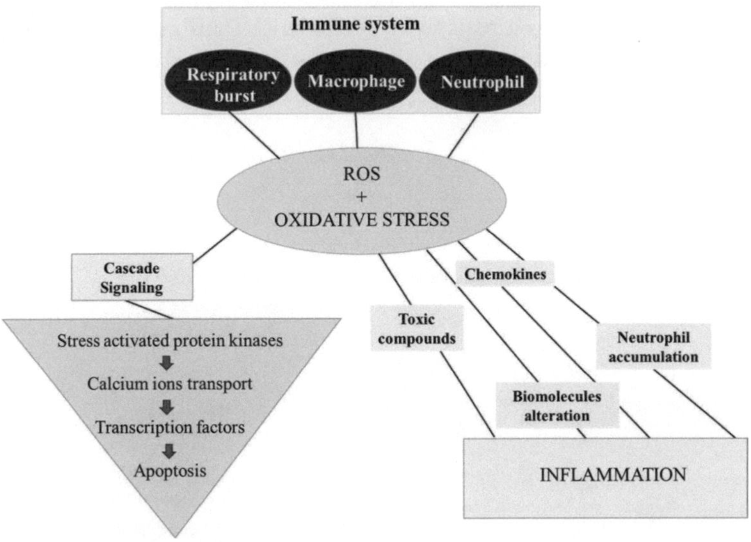


**Figure 1.**
*The source of ROS formation and effects of ROS accumulation [23].*

and secret some pro-inflammatory mediators as a result of their activation. However, upregulation of these pro-inflammatory mediators caused perpetuates the intestine's inflammatory response in these conditions. Therefore, overexpression of pro-inflammatory cytokines plays a crucial role in IBD. However, the types of inflammatory reaction of the immune response are different in CD and UC [30]. Thus, the utilization of anti-inflammatory bioactive compounds is a safe approach for the treatment of IBD by regulating pro-inflammatory mediators.

## 3. Essential oils (EOs)

EOs are volatile compounds that are found in different parts of medicinal and herbal plants. Most of the EOs are known as generally recognized as safe (GRAS) and can be used as food preservatives or flavoring agents. EOs consist of various active constituents with numerous biological properties that are influenced by their chemical diversity and quantities [31]. EOs are more complex and comprise several components at different concentrations. They are defined by some main constituents at relatively high concentrations (> 20%) compared to other components present in trace amounts. EOs are classified into two main groups including terpenes/terpenoids and aromatic/aliphatic compounds. The various biological attributes of EOs are related to their major compounds [32]. In addition, phenolic compounds, plant secondary metabolites which consist of a minimum aromatic ring with at least one or more hydroxyl groups, are sub-classified into two groups: flavonoids and non-flavonoids. Flavonoids are subdivided into many groups comprising flavones, flavan-3-ols, dihydrochalcones, dihydroflavonols, flavonols, flavanones, proanthocyanins, anthocyanins, chalcones, isoflavones, and aurones. Non-flavonoids comprise phenolic acids, stilbenes, lignans, coumarins, curcuminoids, and tannins. The important source of phenolic compounds are nuts, soy products, cocoa, vegetables, cereals, red wine, soy products, whole grains, and olive oil. Cardio-protective and anti-inflammatory traces of polyphenolics have been studied and probably have a positive effect on IBD and IBS [4]. **Figure 2** depicts the chemical structure of some constituents of EOs.

EOs have recently gained increasing attention with their potential biological activates and have been largely used in the pharmaceutical industry as safe and natural

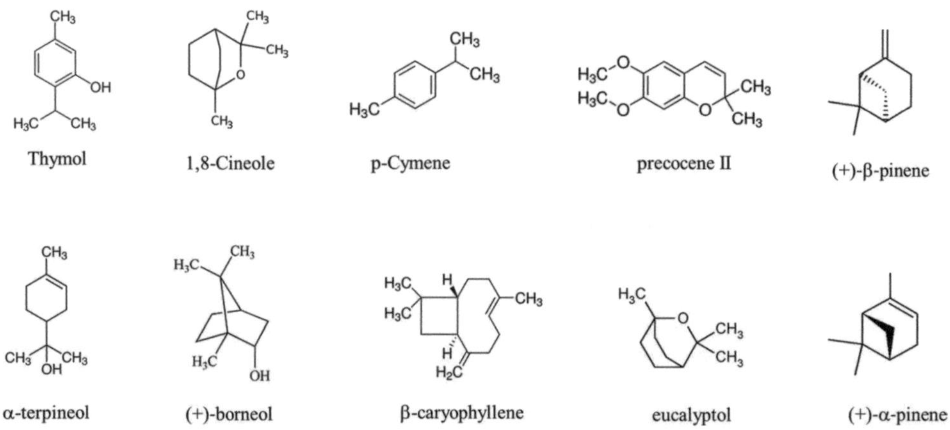


**Figure 2.**
*Some of the main bioactive constituents of EOs.*

medicines. They are well recognized to possess antioxidant, anti-inflammatory, antimicrobial, antiviral, and anticarcinogenic properties [33]. Recently, some investigations confirmed that EOs from various plants have anti-inflammatory activity by reducing or inhibiting of the production of pro-inflammatory mediators [34]. EOs exhibit their anti-inflammatory mechanisms by reducing the gene expression of pro-inflammation cytokines (TNF-α, IL-1β, IL-6 and IFN-γ) and enzymes such as inducible nitric oxide synthase (iNOS), cyclooxygenase (COX-2) and Myeloperoxidase (MPO) which caused upregulation of inflammatory responses.

### 3.1 Essential oils anti-inflammatory activity

Numerous investigations reported that intake of EOs appeared to have protective activity against pro-inflammatory products. For instance, Amorim et al. [35] suggested that *Citrus* species EOs possess anti-inflammatory activity. This study showed that the *C. limon*, *C.latifolia*, *Citrus aurantifolia* and *C. limonia* (10–100 mg/kg, p.o.) EOs caused reducing the cytokines mediators such as TNF-α, interleukin-1β (IL-1β), and interferon-γ (IFN-γ) in carrageenan-induced inflammation in a subcutaneous air pouch (SAP) model. The anti-inflammatory activity of *C. limon* and *C. limonia* might be related to the high amounts of limonene. Another study suggested significantly decreased IL-6 secretion by *Pinus* EOs. They explained that the low concentration (0.01%) of *Pinus heldreichii* Christ (Pinaceae), *Pinus peuce* Griseb and *Pinus mugo* EOs can decrease the IL-6 production up to 60%. The EOs of *Pinus* mainly consist of α-pinene [34]. The chemical structure of EOs from leaves of *Ocimum basilicum*, *Ocimum americanum*, *Hyptis spicigera*, *Lippia multiflora*, *Ageratum conyzoides*, *Eucalyptus camaldulensis* and *Zingiber officinale* was also investigated. α-terpineol (59.78%) and β-caryophyllene (10.54%) for *O. basilicum*; 1, 8-cineole (31.22%) and camphor (12.730%) for *O. americanum*; β-caryophyllene (21%) and α-pinene (20.11%) for *H. spicigera*; p-cymene (25.27%), β-caryophyllene (12.70%), thymol (11.88) for *L. multiflora*; precocene (82.10%) for *A. conyzoides*; eucalyptol (59.55%) and α-pinene (9.17%) for *E. camaldulensis*; arcurcumene (16.67%) and camphene (12.70%) for *Z. officinale* were determined as the main compositions of EOs which have impact on their antioxidant and anti-inflammatory attributes. Among all EOs, *Z. officinale* (0.4 mg/ml) showed the highest anti-inflammatory activity by 50.9% of inhibition of lipoxygenase. The anti-inflammatory activities were investigated according to the prevention effect of lipoxygenase which plays a significant role in several human cancers [36]. A study also investigated the anti-inflammatory attributes of *Thymus vulgaris* EOs. They found that EOs can cease the 5-lipoxygenase activity and lower the TNF-α, IL-1β, and IL-8 secretion in THP-1 cells [37]. Siani et al., [38] also investigated the anti-inflammatory activity of *Syzgium cumini* and *Psidium guajava* EOs in lipopolysaccharide (LPS)-induced pleurisy model. Both types of EOs showed anti-inflammatory properties via prevention influence on eosinophil and, to a lesser extent, neutrophil and mononuclear cell migration. This effect of *Syzygium cumini* and *P. guajava* mainly is contributed to their compositions including limonene, ocimenes, α-pinene, and caryophyllene-type sesquiterpenes. Furthermore, some substances have synergistic effect in combination together such as α-pinene and mono- or sesquiterpenes [38]. Thymus EOs decreased the gene expression of nuclear factor-kappa (NF-k)B, COX-2 and iNOs, consequently causing lower production of NO and TNF-α [39]. The carvacrol and thymol are two main constituents of thyme EOs which are responsible for thyme anti-inflammatory attributes by inhibiting of cyclooxygenase activity and NO production [40]. Similar observations were revealed about the anti-inflammatory properties of *T. caespititius* [41] and *Thymus pulegioides*, *T. praecox* subsp. *polytrichus*, *T. vulgaris*, *Thymus serpyllum* subsp. *serpyllum*, *T. longicaulis*, *T. striatus* extracts [42]. Moreover, EOs isolated from algae have anti-inflammatory activity as Dhara and Chakraborty [43] reported that xenicane-type diterpenoid from *Sargassum ilicifolium* exhibit the inhibitory effect against pro-inflammatory enzymes such as 5-lipoxygenase and COX-2.

## 3.2 Protocols of inflammatory bowel disease treatment by essential oils

There are several studies about the therapeutic and pharmaceutical attributes of EOs related to various diseases [43]. An alternative approach to the treatment of IBDs in the administration of EOs and several investigations are presented in **Table 1**.

| Essential oil | Study design | Major results | Reference |
|---|---|---|---|
| Lavender EO (LEO) | DSS- induced colitis in mice | • Alleviate DSS-induced colonic, adjust the rate of inflammatory factors.<br>• Alleviate UC mice mucosal injury and prevent inflammatory reactions. | Wang, et al. [16] |
| Crude & Bran-Processed *Atractylodes lancea* EO (ALEO) | LPS- induced inflammatory injury of human colonic epithelial cells | • Anti-inflammatory trace of ALEO on LPS-induced HCoEpiC.<br>• Bran-processed AL essential oil was more effective.<br>• Mechanism: IKK/NF-κB signaling pathway. | Yu, et al. [44] |
| *Satureja Khuzestanica Jamzad* EO (SKEO) | Acetic acid-induced colitis in mice | • SKEO possesses antioxidant, antimicrobial, anti-inflammatory, and antispasmodic effect.<br>• Protects animals against experimentally induced IBD. | Ghazanfari, et al. [45] |
| Hydroalcoholic extract of *Tagetes minuta L.* | *In vitro* study in human Peripheral Blood Mononuclear Cells | • The two pheophytins inhibited the production of NF-Kb.<br>• They have the anti-inflammatory activity of the Huacatay extracts and their use in the treatment of stomach and intestine discomfort.<br>• Aqueous and hydroalcoholic extracts possessed anti-inflammatory activity in vitro.<br>• The hydroalcoholic extract was the most active (IC50 between 59.72 and 66.42 μg/mL) in all cell lines. | Ticona, et al. [46] |
| *Rosmarinus officinalis L.* EO (ROEO) | TNBS-induced colitis in rats | • Both the RHE and ROEO had an anticolitic effect.<br>• Can be applied for remedy of inflammatory bowel diseases in traditional medicine.<br>• Alpha-pinene had an inhibitory effect on the nuclear translocation of factor-kappa B (NF-kappa B). | Minaiyan, et al. [47] |

| Essential oil | Study design | Major results | Reference |
|---|---|---|---|
| Cinnamon EO (CEO) | DSS-induced colitis in mice | • Improving the intestinal flora imbalance by the inhibitory trace of CEO on IBD.<br>• TLR4 and TNF-α were positively correlated with Helicobactermodified. | Li, et al. [48] |
| *Carum carvi L.* EO (CCEO) | TNBS-induced colitis in rats | • CCEO (hydroalcoholic extract and EO) own anti-colitis activity. | Keshavarz, et al. [49] |
| *Ocimum basilicum L.* EO (OBEO) | Acetic acid-induced colitis in rats | • 200 and 400 μL/kg of OBEO decreased the enhancement of myeloperoxidase.<br>• OBEO possessed anti acetic acid-induced colitis effect. | Rashidian, et al. [50] |
| *R. officinalis* EO (ROEO) | TNBS- induced colitis in mouse model | • ROEO is able to influence several variables of murine experimental inflammatory models depending on the concentration used.<br>• The anti-inflammatory effects of ROEO should be interpreted carefully due to its time and dose-related effects. | Juhás, et al. [51] |
| *Pelargonium graveolens* EO (PGEO) | Acetic acid-induced UC in rats | • Significantly lower score values of macroscopic and microscopic characters when compared to the acetic acid-treated group.<br>• Deproherb® inhibited the acetic acid toxic reactions in the rat bowel. | Bastani, et al. [52] |
| *Origanum onites L.* EO | TNBS-induced colitis in the rats | • Significant protective effect on the colonic injury. | Dundar, et al. [53] |
| Limonene from *Agastache mexicana* | Oxazolone-induced colitis in mice | • Antioxidant and anti-inflammatory effect of limonene.<br>• Limonene diminish signaling pathway of iNOS, COX-2, PGE2, TGF-β, and ERK1/2.<br>• Agastache mexicana EO impedes intestinal tissue damage.<br>• Agastache mexicana EO diminishes the myeloperoxidase activity.<br>• Prevention of cytokines such as IL-1, IL-6, TNF-α, and INF-γ expression<br>• Lessening pain in the UC in humans. | Estrella, et al. [8] |
| *Foeniculum vulgare* EO (FVEO) | Acetic acid-induced colitis in rat | • Diminish the macroscopic and microscopic injuries compared to the acetic acid group.<br>• Diminish the MPO activity and the TNF-α positive cells expression in the colon tissue contrasted to the acetic acid group.<br>• Prevent expression of p-NF-kB p65 protein induced by acetic acid.<br>• Anti-inflammatory activity effect on colitis induced by acetic acid in rats which prevents the NF-kB pathway. | Rezayat, et al. [54] |

| Essential oil | Study design | Major results | Reference |
|---|---|---|---|
| *Zanthoxylum bungeanum* Pericarp EO (ZBEO) | DSS-induced colitis in mice | • ZBEO increased levels of the commensal bacteria containing *Lactobacillus* and *Bifidobacteria.*<br>• ZBEO reduced *E. coli* levels in the feces of mice.<br>• Supplementation with ZBEO might provide a new dietary strategy for the prevention of UC. | Zhang, et al. [55] |
| Thyme EO (TEO) | TNBS- induced colitis in mice | • Caused a significant inhibition of total mRNA IL-1β expression in the - mouse colon.<br>• Decreased the macroscopic and microscopic scores of colitis.<br>• In 1250 ppm concentration in the diet increased ear edema induced by oxazolone application in mice.<br>• Can affect murine experimental inflammatory models depending on the concentration used. | Juhás, et al. [56] |
| Garden thyme EO (GTEO) | Intestinal inflammatory status of rainbow trout (*Oncorhynchus mykiss*) | Dietary GTEO supplementation:<br>• Ameliorated the increased TNF-α.<br>• Transforming growth factor-β and interleukin-8 expression induced by dietary AFB1 contamination.<br>• Significantly enhanced interleukin-1β expression. | Ghafarifarsani, et al. [57] |

**Table 1.**
*Treatment of inflammatory bowel disease by essential oils.*

Yu, et al. [44] investigated the anti-inflammatory effect of *Atractylodes lancea* EOs against UC *in vitro*. They proved that Atractylodes lancea EOs can downregulate the level of IL-6, IL-8, IL-12, IL-1-β, TNF-α, NO, p-IKK-α, p-IKK-β, and NF-κB human colonic epithelial (HcoEpiC) cells induced by LPS- epithelial cells. IKK/NF-κB signaling pathway was the *in vitro* mechanism.

In *vivo* experimental studies have shown the therapeutic effects of EOs on UC inflammation. In 2016, Rashidian, et al. [50] conducted a study of the meliorative effect of *O. basilicum L.* EO after two doses in acetic acid-induced rat model. The results showed that the treatment with 200 and 400 μL/kg of EO caused a significantly reduction in the ulcer severity, ulcer area, and ulcer index and confirmed the protective activity of EOs. Moreover, *Lavender* EO has also been shown to improve colonic mucosal injury in dextran sulfate sodium (DSS)-induced UC mice by reducing the inflammatory cytokines levels such as in serum and colon tissue's EGFR, TNF-α, and IFN-γ. The key pathway in the UC treatment is the Th17 cell differentiation, PI3K-Akt signaling pathway, and Th1 and Th2 cell differentiation of lavender EOs [16]. Estrella, et al. [8] also demonstrated that Limonene from *Agastache mexicana* EO has a potential effect on improving the UC in Swiss Webster mice. They investigated the *A. mexicana* ssp. *mexicana* EO (3–300 mg/kg) activity in an oxazolone-induced colitis model. According to the results, limonene possessed antioxidant and anti-inflammatory bioactivity that caused downregulation of the iNOS, COX-2, PGE2, TGF-β, and ERK1/2 signaling pathway. Thus, EOs prevented intestinal tissue damage and reduced myeloperoxidase activity, macroscopic damage reducing and inhibition of cytokines expressions such as IL-1, IL-6, TNF-α, and INF-γ. Moreover, EOs have antinociceptive attributes resulting in lower pain in the UC in human. It has been suggested that 200 and 400 mg/kg of *F. vulgare* EOs decrement the TNF-α positive cells expression of colon tissue. They have a considerable effect on rat inflammatory of acetic acid-induced colitis by the prohibition of NF-kB pathway [54]. Furthermore, it has been reported that both *Carum carvi L.* (caraway) EOs and hydroalcoholic extract own anti-inflammatory properties in colitis induced by trinitrobenzene sulfonic acid (TNBS) in rats. The ulcerative lesion index would be prevented by 100–400 μl/kg orally administration of *C. carvi L.* EOs. The inflammatory cytokines and chemokines can be reduced by the caraway terpenoid, flavonoids, fatty acids, triacylglycerols, polysaccharides, lignin, and polyacetylenic compounds, resembling the glucocorticoids mechanism. It seems that caraway reduces the production of prostaglandin E2 and increases the production of leukotriene B4 in human polymorphonuclear leucocytes [49]. A similar observation was reported by Minaiyan, et al. [47] who study the anti-colitis activity of *Rosmarinus officinalis* L. EOs (100, 200, and 400 μl/kg) and extract (100, 200, and 400 mg/kg) in rats colitis induced by TNBS. Moreover, 100–400 mg/kg of *Pelargonium graveolens* EOs own dose-independent anti-inflammatory potential in acetic acid-induced rat UC induced by acetic acid thus *P. graveolens* EOs diminish the oxidative stress by inhibiting the production of free radicals, and in the end preventing the increase of inflammation [52].

By adjustment of intestinal microflora, EOs effect on IBD. For example, *cinnamon* EOs administration amends the diversity and richness of the intestinal microbiota, reduces in *Helicobacter* and *Bacteroides,* and increase in Bacteroidales_S24–7 family in mouse colitis induced by DSS. There is a positive correlation between TNF-α with *Helicobacter.* It seems that the protective attributes of *cinnamon* EOs against IBD is attributed to cinnamaldehyde [48]. EOs of *R. officinalis* also is full of terpenes and 1.8-cineole is the main compound of its EO by antinociceptive and anti-inflammatory trace. This terpene remarkably prevents the production of cytokines in lymphocytes and monocytes [51].

## 4. Conclusion

EOs are a blend of volatile, aromatic, and natural substances extracted as secondary metabolites from different parts of plants or algae. EOs have a great potential to be used as a part of pharmaceuticals, nutraceuticals, and functional foods because of their broad range of biological activities. In the recent years, EOs have gained much attention due to their antioxidant and anti-inflammatory attributes. The present chapter revealed encouraging results about numerous EOs being used in different IBD and UC animal models. IBD and its attributed disorders such as CD and UC dramatically increase in recent years because of various reasons such as age, genetics, immune response, and lifestyle, in a particular diet. While several studies confirmed that the EOs exhibit anti-inflammatory effects via in downregulation of gene expression of cytokines pro-inflammatory and related enzymes. This chapter suggests the utilization of EOs as healthy food ingredients or dietary supplements with anti-inflammatory characteristics.

## Conflict of interest

The authors declared no conflict of interest.

## Author details

Najmeh Oliyaei[1], Nader Tanideh[1,2]* and Seyedeh Zahra Nasirifar[3]

1 Stem Cells Technology Research Center, Shiraz University of Medical Sciences, Shiraz, Iran

2 Department of Pharmacology, School of Medicine, Shiraz University of Medical Sciences, Shiraz, Iran

3 Faculty of Nutrition Sciences and Food Technology, Department of Food Science and Technology, Shahid Beheshti University of Medical Sciences, Tehran, Iran

*Address all correspondence to: tanidehn@gmail.com

## References

[1] Pérez GS et al. Anti-inflammatory activity of some essential oils. Journal of Essential Oil Research. 2011;**23**(5):38-44

[2] Zhou X-L et al. Fecal lactoferrin in discriminating inflammatory bowel disease from irritable bowel syndrome: A diagnostic meta-analysis. BMC Gastroenterology. 2014;**14**(1):1-8

[3] Ma Q et al. A review of pharmacological and clinical studies on the application of Shenling Baizhu san in treatment of ulcerative colitis. Journal of Ethnopharmacology. 2019;**244**:112105

[4] Hagan M, Hayee BH, Rodriguez-Mateos A. (poly) phenols in inflammatory bowel disease and irritable bowel syndrome: A review. Molecules. 2021;**26**(7):1843

[5] Guan Q, Zhang J. Recent advances: The imbalance of cytokines in the pathogenesis of inflammatory bowel disease. Mediators of Inflammation. 2017;**2017**:1-8

[6] Yang C et al. Algal oil rich in docosahexaenoic acid alleviates intestinal inflammation induced by antibiotics associated with the modulation of the gut microbiome and metabolome. Journal of Agricultural and Food Chemistry. 2021;**69**(32):9124-9136

[7] Fan W et al. Genistein-derived ROS-responsive nanoparticles relieve colitis by regulating mucosal homeostasis. ACS Applied Materials & Interfaces. 2021;**13**(34):40249-40266

[8] Estrella G-RA et al. Limonene from Agastache mexicana essential oil produces antinociceptive effects, gastrointestinal protection and improves experimental ulcerative colitis. Journal of Ethnopharmacology. 2021;**280**:114462

[9] Naini MA et al. Anti-inflammatory, antioxidant, and healing-promoting effects of Aloe vera extract in the experimental colitis in rats. Evidence-based Complementary and Alternative Medicine. 2021;**2021**:1-12

[10] Tanideh N et al. Evaluating the effect of melillotus officinalis L. aqueous extracts on healing of acetic acid-induced ulcerative colitis in male rats. Annals of Colorectal Research. 2016;**4**(4):1-17

[11] Lombardo GE et al. Mechanisms underlying the anti-inflammatory activity of bergamot essential oil and its antinociceptive effects. Plants. 2020;**9**(6):704

[12] Patil MVK, Kandhare AD, Bhise SD. Effect of aqueous extract of Cucumis sativus Linn. Fruit in ulcerative colitis in laboratory animals. Asian Pacific Journal of Tropical Biomedicine. 2012;**2**(2):S962-S969

[13] Tanideh N et al. Effects of hydroalcoholic extract of Ziziphus jujuba on acetic acid induced ulcerative colitis in male rat (Rattus norvegicus). Journal of Coloproctology. 2016;**36**:189-195

[14] Almeer RS et al. Ziziphus spina-christi fruit extract suppresses oxidative stress and p38 MAPK expression in ulcerative colitis in rats via induction of Nrf2 and HO-1 expression. Food and Chemical Toxicology. 2018;**115**:49-62

[15] Zhang Y et al. The preventive effect and underlying mechanism of Rhus chinensis mill. Fruits on dextran sulphate sodium-induced ulcerative colitis in mice. Food & Function. 2021;**12**(20):9965-9978

[16] Wang Y et al. The mechanism of lavender essential oil in the

treatment of acute colitis based on "quantity–effect" weight coefficient network pharmacology. Frontiers in Pharmacology. 2021;**12**:1-13

[17] Hanif MA et al. Essential oils. In: Essential Oil Research. Switzerland: Springer Nature; 2019. pp. 3-17

[18] Atasever-Arslan B et al. Screening of new antileukemic agents from essential oils of algae extracts and computational modeling of their interactions with intracellular signaling nodes. European Journal of Pharmaceutical Sciences. 2016;**83**:120-131

[19] Medrano JJA. Essential oils in the development of new medicinal products. In: Essential Oils-Oils of Nature. Rijeka: IntechOpen; 2019

[20] Nazzaro F et al. Essential oils and microbial communication. In: Essential Oils-Oils of Nature. United Kingdom: IntechOpen; 2019. pp. 165-190

[21] Mehdizadeh L, Moghaddam M. Essential oils: Biological activity and therapeutic potential. In: Therapeutic, Probiotic, and Unconventional Foods. Elsevier; 2018. pp. 167-179

[22] Song Y et al. Biological functions of NLRP3 Inflammasome: A therapeutic target in inflammatory bowel disease. Cytokine & Growth Factor Reviews. 2021;**60**:61-75

[23] Balmus IM et al. The implications of oxidative stress and antioxidant therapies in inflammatory bowel disease: Clinical aspects and animal models. Saudi Journal of Gastroenterology : Official Journal of the Saudi Gastroenterology Association. 2016;**22**(1):3

[24] Rezaie A, Parker RD, Abdollahi M. Oxidative stress and pathogenesis of inflammatory bowel disease: An epiphenomenon or the cause? Digestive Diseases and Sciences. 2007;**52**(9):2015-2021

[25] Ferreira ICdV. Pharmacological screening of essential oils for identification of lead compounds to target inflammatory bowel disease; 2011

[26] Vecchi Brumatti L et al. Curcumin and inflammatory bowel disease: Potential and limits of innovative treatments. Molecules. 2014;**19**(12):21127-21153

[27] Tanideh N et al. Healing effect of pistacia atlantica fruit oil extract in acetic acid-induced colitis in rats. Iranian Journal of Medical Sciences. 2014;**39**(6):522

[28] Hosseinzadeh F et al. The healing effect of grape seed oil enema with or without sesame oil in acetic acid induced ulcerative colitis of rats. World Journal of Plastic Surgery. 2017;**6**(2):176

[29] Tanideh N et al. Protection by pure and genistein fortified extra virgin olive oil, canola oil, and rice bran oil against acetic acid-induced ulcerative colitis in rats. Food & Function. 2020;**11**(1):860-870

[30] Gálvez J. Role of Th17 cells in the pathogenesis of human IBD. International Scholarly Research Notices. 2014;**2014**:1-14

[31] Tanveer M et al. Spicing up gastrointestinal health with dietary essential oils. Phytochemistry Reviews. 2020;**19**(2):243-263

[32] Kar S, Gupta P, Gupta J. Essential oils: Biological activity beyond aromatherapy. Natural Product Sciences. 2018;**24**(3):139-147

[33] Shaaban HA, El-Ghorab AH, Shibamoto T. Bioactivity of essential oils

and their volatile aroma components. Journal of Essential Oil Research. 2012;**24**(2):203-212

[34] Basholli-Salihu M et al. Phytochemical composition, anti-inflammatory activity and cytotoxic effects of essential oils from three Pinus spp. Pharmaceutical Biology. 2017;**55**(1):1553-1560

[35] Amorim JL et al. Anti-inflammatory properties and chemical characterization of the essential oils of four citrus species. PLoS One. 2016;**11**(4):e0153643

[36] Bayala B et al. Chemical composition, antioxidant, anti-inflammatory and anti-proliferative activities of essential oils of plants from Burkina Faso. PLoS One. 2014;**9**(3):e92122

[37] Tsai M-L et al. Antimicrobial, antioxidant, and anti-inflammatory activities of essential oils from five selected herbs. Bioscience, Biotechnology, and Biochemistry. 2011;**75**(10):110377-1-7

[38] Siani AC et al. Anti-inflammatory activity of essential oils from Syzygium cumini and Psidium guajava. Pharmaceutical Biology. 2013;**51**(7):881-887

[39] Lorenzo JM et al. Understanding the potential benefits of thyme and its derived products for food industry and consumer health: From extraction of value-added compounds to the evaluation of bioaccessibility, bioavailability, anti-inflammatory, and antimicrobial activities. Critical Reviews in Food Science and Nutrition. 2019;**59**(18):2879-2895

[40] Fachini-Queiroz FC et al. Effects of thymol and carvacrol, constituents of Thymus vulgaris L. essential oil, on the inflammatory response. Evidence-based Complementary and Alternative Medicine. 2012;**2012**:1-10

[41] Afonso AF et al. Health-promoting effects of Thymus herba-barona, Thymus pseudolanuginosus, and Thymus caespititius decoctions. International Journal of Molecular Sciences. 2017;**18**(9):1879

[42] Kindl M et al. Antioxidant and anticholinesterase potential of six thymus species. Evidence-based Complementary and Alternative Medicine. 2015;**2015**:1-10

[43] Dhara S, Chakraborty K. Anti-inflammatory xenicane-type diterpenoid from the intertidal brown seaweed Sargassum ilicifolium. Natural Product Research. 2020;**35**(24):5699-5709

[44] Yu Y et al. Comparison of the effects of essential oil obtained from the crude and bran-processed Atractylodes lancea on lipopolysaccharide-induced inflammatory injury of human colonic epithelial cells by downregulating the IKK/NF-κB Signaling pathway. Evidence-based Complementary and Alternative Medicine. 2021;**2021**:1-16

[45] Ghazanfari G et al. Biochemical and histopathological evidences for beneficial effects of Satureja Khuzestanica Jamzad essential oil on the mouse model of inflammatory bowel diseases. Toxicology Mechanisms and Methods. 2006;**16**(7):365-372

[46] Ticona LA et al. Hydroalcoholic extract of Tagetes minuta L. inhibits inflammatory bowel disease through the activity of pheophytins on the NF-κB signalling pathway. Journal of Ethnopharmacology. 2021;**268**:113603

[47] Minaiyan M et al. Effects of extract and essential oil of Rosmarinus officinalis L. on TNBS-induced colitis in rats. Research. Pharmaceutical Sciences. 2011;**6**(1):13

[48] Li Al et al. Effect of cinnamon essential oil on gut microbiota in the mouse model of dextran sodium sulfate-induced colitis. Microbiology and Immunology. 2020;**64**(1):23-32

[49] Keshavarz A et al. Effects of Carum carvi L.(caraway) extract and essential oil on TNBS-induced colitis in rats. Research in Pharmaceutical Sciences. 2013;**8**(1):1

[50] Rashidian A et al. Protective effect of Ocimum basilicum essential oil against acetic acid–induced colitis in rats. Evidence-Based Complementary and Alternative Medicine. 2016;**21**(4):NP36-NP42

[51] Juhás Š et al. Anti-inflammatory effects of Rosmarinus officinalis essential oil in mice. Acta Veterinaria Brno. 2009;**78**(1):121-127

[52] Bastani M et al. Biochemical and histopathological evidence for beneficial effects of Pelargonium graveolens essential oil on the rat model of inflammatory bowel disease. Research Journal of Pharmacognosy. 2019;**6**(2):77-84

[53] Dundar E et al. The effects of intra-rectal and intra-peritoneal application of Origanum onites L. essential oil on 2, 4, 6-trinitrobenzenesulfonic acid-induced colitis in the rat. Experimental and Toxicologic Pathology. 2008;**59**(6):399-408

[54] Rezayat SM et al. Foeniculum vulgare essential oil ameliorates acetic acid-induced colitis in rats through the inhibition of NF-kB pathway. Inflammopharmacology. 2018;**26**(3):851-859

[55] Zhang Z et al. In vivo study of the efficacy of the essential oil of Zanthoxylum bungeanum pericarp in dextran sulfate sodium-induced murine experimental colitis. Journal of Agricultural and Food Chemistry. 2017;**65**(16):3311-3319

[56] Juhás Š et al. Anti-inflammatory effects of thyme essential oil in mice. Acta Veterinaria Brno. 2008;**77**(3):327-334

[57] Ghafarifarsani H, Kachuei R, Imani A. Dietary supplementation of garden thyme essential oil ameliorated the deteriorative effects of aflatoxin B1 on growth performance and intestinal inflammatory status of rainbow trout (Oncorhynchus mykiss). Aquaculture. 2021;**531**:735928

Chapter 9

# *Teucrium ramosissimum* Derived-Natural Products and Its Potent Effect in Alleviating the Pathological Kidney Damage in LPS-Induced Mice

*Fatma Guesmi and Ahmed Landoulsi*

## Abstract

*Teucrium* esssential oil mediates an extensive spectrum of biological effects, including renal diseases. The aim of this research was to explore the ethnobotanical feature, biochemical composition and antiinflammatory potential of *T. ramosissimum* alone or prior the use of LPS-induced kidney damage. The essential oils were subjected to Gas chromatography-mass spectrometry (GC/MS) apparatus to detect biomolecules in *T. ramosissimum*. *In vivo* renal dysfunction induced by LPS was investigated using mouse model. Our data showed that oral treatment of animals with LPS highly increased level of serum biomarkers and induces renal dysfuntion, whereas, pre-treatment with *T. ramosissimum* mediated markedly histopatological changes of kidney architecture and ameliorates renal function. Dense cover of secretory structures in *teucrium* leaves may protect this specie. Overall, this study showed phytocompounds richness and interesting biological activities of Tunisian *Teucrium ramosissimum*. Essential oil of this specie *T. ramossimum* given prior to LPS exposure protected mice from renal inflammation.

**Keywords:** *Teucrium ramosissimum*, essential oil, hairs, LPS, renal dysfunction

## 1. Introduction

The genus *Teucrium L.* (Lamiaceae) is a genus growing in mild climate zones, particularly in the Mediterranean Basin and Central Asia [1]. Limited number of in-depth scientific researches have been done so far on the phytochemistry and bioactivities [2] of *Teucrium* genus that is represented by herbs or shrubs, with tubular or campanulate calyx, 2-lipped or actinomorphic, 5-toothed, the teeth equal or the upper larger; corolla with one 5-lobed lip; tube without a ring of hairs inside, often included in the calyx; and nutlets smooth or reticulate [1].

*Teucrium* species have been used in phytopharmacology, helping to treat many diseases, including tuberculosis, gastrointestinal disorders, inflammations, rheumatism, and diabetes [3]. The widespread applications of *Teucrium* genus in the ethnomedicine of several countries [1] may be due to its richness in biocompounds used in *in vivo* and *in vitro* biological effects. Moreover, traditional health care systems based on plants and plant-derived products are highly popular and employed therapeutically [2] in Tunisia. The plant essence that contains several secondary metabolites is synthetized in all parts (leaves, flowers, stem, seeds, buds). The use of essential oils in industries are markedly increased, including the beverage, food, aromatherapy, cosmetics and personal care [4]. Moreover, the recent trend in the field of inflammation research is to search for alternative therapeutic agents from natural sources that are devoid of the adverse effects characteristic for conventional steroids or nonsteroidal anti-inflammatory drugs (NSAIDs). In this context, phytochemical studies and biochemical investigations on the mode of action of traditional complementary remedies are of utmost importance [2]. Acute kidney damage, a great public health problem, has been grown in the world. It's a critical care syndrome and an abrupt loss in renal function [5], resulting in acute reduction of renal activity and up to 22% mortality of hospitalized patients. Acute kidney injury is estimated to occur in about 20–200 per million population in the community, 7–18% of patients in hospital, and approximately 50% of patients admitted to the intensive care unit (ICU) [6]. Lipopolysaccharide (*Escherichia coli* 055:B5) is one of the most important causes of sepsis and is involved in the pathogenesis of sepsis-associated acute kidney injury (SA-AKI), which may lead to "cytokine storm," intensified oxidative stress, low blood pressure, renal hypoperfusion, and finally a gradual decline in renal function [7].

*T. ramosissimum* belongs to the Lamiaceae family, of the genus *Teucrium* is known as "Hchichet Belgacem" or "Hchichet Ben Salem" in the region of Gafsa in the southwest of Tunisia [8]. The specie is present in the South of Tunisia in particular in Djebel Orbata (Zannouch-Gafsa, Tunisia) and Bou Hedma Mount (Sidi Bouzid-Tunisia). It exists as a small sub-shrub, bushy, of silver gray, 8–15 cm tall. The stems are slender, erect and small. While, the leaves are white, with rounded limb with 7 deep crenellations. The inflorescence is pauciflor; the white calyx is 4 mm long, with long acute and sub-natural teeth [8]. Three sesquiterpenoids (teucmosin, 4α-hydroxy-homalomenol C, 1β,4β,7α-trihydroxy-8,9-eudesmene), five sesquiterpenoids (oplopanone, homalomenol C, oxo-*T*-cadinol, 1β,4β,6β-trihydroxyeudesmane, 1β,4β,7α-trihydroxyeudesmane) and two trinorsesquiterpenoids (4β-hydroxy-11,12,13-trinor-5-eudesmen-1,7-dione and 1β,4β-dihydroxy-11,12,13-trinor-8,9-eudesmen-7-one) were isolated from the ethanolic extracts of the aerial parts of *Teucrium ramosissimum* [9]. *Teucrium ramosissimum* is particularly present in the higher mounts of southern Tunisia. The ethnopharmacological uses of this specie in Tunisia are for treatment of inflammation. In fact, many people apply the powder of this species on the external inflamed area to reduce swelling and pain.

The present study provides a new insight into the organ architecture of *Teucrium ramosissimum* Desf. Biochemical compounds of *T. ramosissimum* leaves were detected using GC/MS apparatus. Histological analysis and enzyme levels showed that *T. ramosissimum* decreased LPS-mediated acute kidney injury by inhibiting tissues inflammation and reducing kidney tissue damage.

## 2. Materials and methods

### 2.1 Chemicals

*T. ramosissimum* essential oil solutions (100 mg/ml) were diluted in dimethyl sulfoxide for *in vivo* analysis. 5-Fluorouracil (5-FU) and LPS (*Escherichia coli* 055:B5) was obtained from Sigma-Aldrich Chemicals Co. (St. Louis, MO, USA).

## 2.2 Plant materials

Leaves of *T. ramosissimum* were collected from the mount of Orbata (Gafsa, Tunisia) during the Springer (2018). Essential oil was extracted by Clevenger apparatus. Characterization of phytochemicals by GC-MS analysis indicated the presence of mono- and sesquiterpenic compounds.

## 2.3 Analyses of oily fractions of *T. ramosissimum* with GC/MS

Oily fractions (diluted in 10% hexane) were analysed using GC/MS on a model 6890 gas chromatograph with an autosampler coupled with an Agilent 5973 Mass Selective *Detector* (Agilent Technologies, Palo Alto, CA, USA) with an electron impact ionization of 70 eV. A Phenomenex capillary column, ZB-5MSi (30 m × 250 μm i.d., 0.5 μm film thickness) (Agilent Technologies, Hewlett-Packard, CA, USA) at a temperature rising from 40 to 280 °C (5 °C/min). The carrier gas at a purity of 99.999% used for GC/MS analyses was helium at a flow rate of 0.7 ml/min, a scan time of 1 s and mass range *m/z* 50–550. The terpenic compounds were identified by matching their retention indices with those of the Wiley 09 NIST 2011 mass spectral library of the apparatus.

## 2.4 Protective effects of *T. ramosissimum* against LPS-induced renal inflammation

### *2.4.1 Experimental design*

Both sexes of Swiss albino mice (48, 25 g weight) were divided into 8 groups (n=6) and maintained in plastic cages (polypropylene). Mice, provided from Pasteur Animal Laboratory (Tunisia, Ethic# LNSP/Pro 152012), were housed under animal conditions (25 ± 5°C; 45–55 % relative humidity; 12 h light/dark cycles with free access to water and food.

Group 1: Normal control, orally treated with saline;

Group 2: negative control, orally treated with 10 μg/ml LPS;

Group 3: orally treated with 20 μg/kg *T. ramosissimum* essential oil diluted in Tween 80 (2%);

Group 4: orally treated with 50 μg/kg *T. ramosissimum* essential oil diluted in Tween 80 (2%);

Group 5: comparator control, orally treated with 20 mg/ kg/day 5-FU;

Group 6: orally treated with the mixture of LPS and *Teucrium* essential oil (10 μg/ml and 20 μg/kg, respectively);

Group 7: orally treated with the mixture of LPS and *Teucrium* essential oil (10 μg/ml and 50 μg/kg, respectively);

Group 8: orally treated with the mixture of LPS and 5-FU (10 μg/ml and 20 mg/kg, respectively).

Animals received drugs for one week. In the groups 6, 7 and 8, mice were treated with *Teucrium* essential oil or 5-FU 1 hour before LPS administration.

At the 8th day, mice were sacrificed and blood samples were collected by glass capillary tubes for plasma biomarker analysis. Kidney tissues were collected and processed for microscopic analysis.

### 2.5 Statistical analyses

Statistical analyses of *in vivo* study were performed using GraphPad Prism 4.00 to compare different groups with each other we used a two-way analysis of variance (ANOVA), followed by Tukey's multiple test. A value of $P < 0.05$ was considered statistically significant.

## 3. Results and discussion

### 3.1 Ethnobotanical and phytochemical analysis of *T. ramosissimum*

*T. ramosissimum* macromorphological features (nutlets, leaves, stems, flowers) are shown in **Figure 1**. The major phytocompounds were β-phellandrene, α-cadinol, T-Muurolol, α-bisabolol, camphor, endo-borneol, and epi-bicyclosesquiphellandrene (**Figure 2**).

### 3.2 Effects of *T. ramosissimum* on histological changes of cecum and serum biomarkers of the kidneys

LPS mediated a significantly ($P<0.05$) increase in the levels of plasma urea, creatinine and uric acid (**Figure 3A**). The pretreatment of mice with *T. ramosissimum* notably reduced the levels of plasma biomarkers in serum. Macroscopic features of kidney taken from treated and untreated groups are shown in **Figure 3Bi**. As depicted in **Figure 3Bii**, normal glomerular histoarchitectural and numerous tubules was seen in the kidney and a significant increase in the tubular injury scores after LPS treatment, otherwise, we noted fibrotic lesions, leukocyte infiltration indicative of the inflammation within different areas in the glomeruli, and renal tubules degeneration associated to tubular epithelium desquamation, while no observed damage was detected in mice from normal control group. After one week of treatment of mice with LPS, pronounced tubular necrosis and kidney fibrotic scaring was observed in kidney, which was significantly reversed by *T. ramosissimum* treatment that improves significantly the renal functions when it was given prior to LPS administration (**Figure 3Biii**).

**Figure 1.**
T. ramosissimum *plant. a: habitus; a: leaves; b,c: leaves; d,e: seeds.*

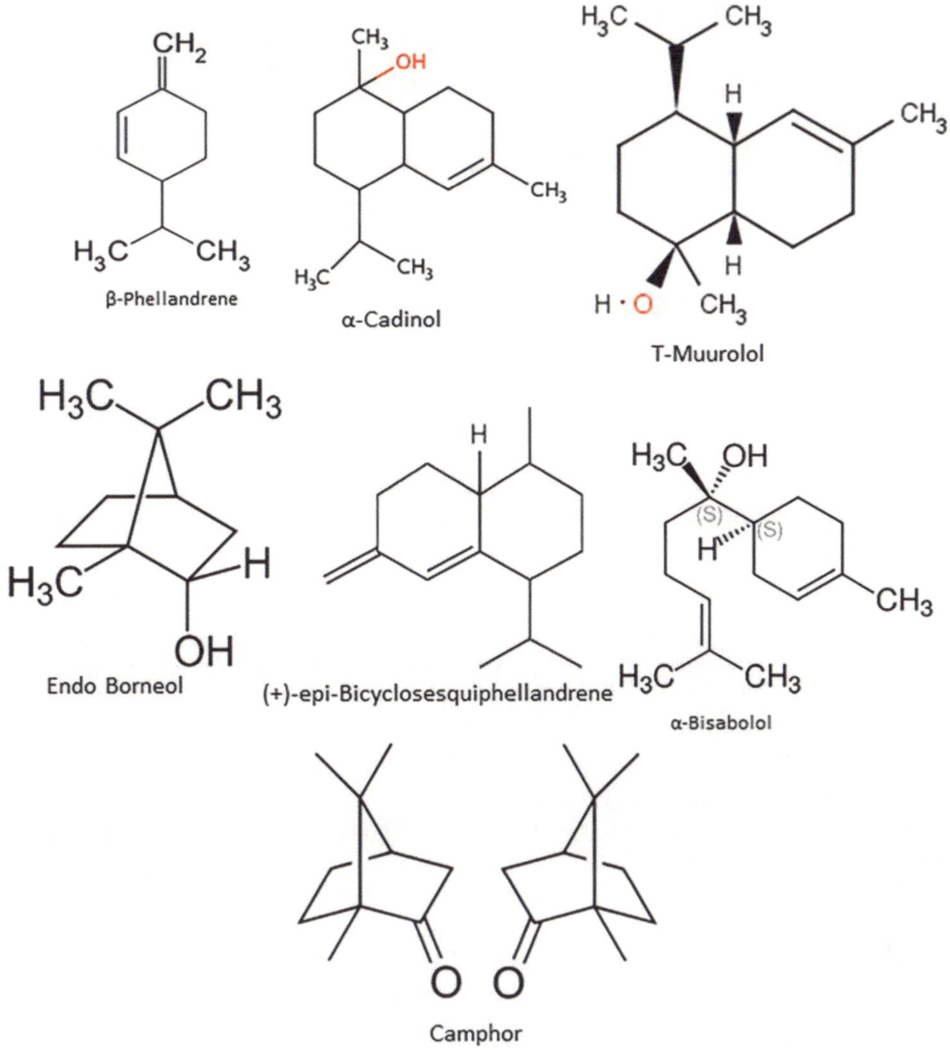


**Figure 2.**
*Phytocompounds of* T. ramosissimum *essential oil isolated using GC/MS analysis.*

Free radicals- induced lipid peroxidation to be one of the major causes of cell membrane damage resulting in a series of pathological situations by causing acute and chronic renal injuries [10]. In fact, LPS is one of the most important causes of sepsis and is involved in the pathogenesis of SA-AKI, which may lead to "cytokine storm," intensified oxidative stress, low blood pressure, renal hypoperfusion, and finally a gradual decline in renal function [7]. In this report, photomicrograph overview revealed the potent protective effect of *T. ramosissimum* that could effectively attenuate the pathological cecum and renal alteration in LPS-induced colorectal inflammation and acute kidney injury in mice model by reducing inflammation. These imply that *T.ramosissimum* pretreatment may attenuates the pathological features of LPS-induced inflammation revealed by Hematoxylin & Eosin (H&E) staining.

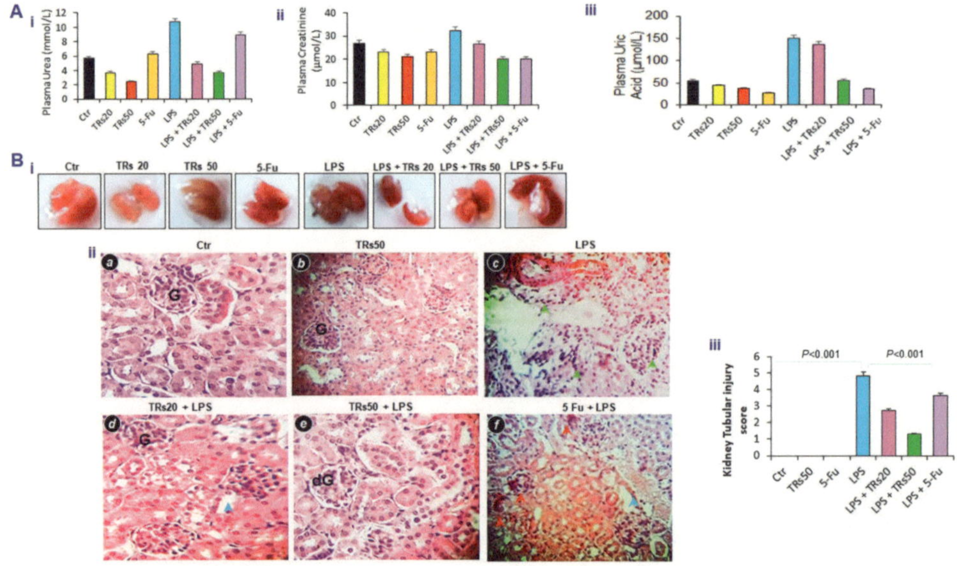

**Figure 3.**
*A. Creatinine (i), urea (ii) and UA (iii) levels in plasma. B. Macroscopical view of mice kidney (i), histopathological analysis (ii) of kidney sections (scale bar = 250 μm) observed by H&E staining and Kidney tubular injury score (iii) of treated groups. Green arrows- inflammatory cell infiltration; red arrows: hemorage; blue arrows: edema of the intertubular spaces; TRs:* Teucrium ramosissimum*; G: glomerulus; dG: degenerated glomerulus. (H&E staining, Magnification: a–e ×40; f ×10). Data represent mean ± SD, n=6; ** P<0.01 vs. LPS.*

The current results further support previous findings on the effect of LPS to mediate tubule and glomerulus degeneration.

*T. ramosissimum* given prior to LPS treatment induced decrease in serum biomarkers (urea, uric acid, creatinine). In this report, plasma creatinine and urea increased significantly in LPS-treated group, and this indicate diminished ability of the kidneys to filter these waste products from the blood and excrete them in the urine [11]. Additionally, this work demonstrated that LPS increased uric acid that mediated arteriolopathy and interstitial inflammation suggest mechanisms that would exacerbate or potentiate progressive renal functional decline after injury. This process involves accumulation of free radicals. Moreover, in the kidney, LPS binds to TLR4 proteins and mediates the proinflammatory cytokines release, and more precisely IL-1β and TNF-α [7] and induces the transcriptional factor, NF-kB, activation that regulates a variety of inflammatory gene expression [12]. Effectively, any compound able to modulate inflammation or inflammation-related processes can be thought of as a renal protective agent and/or a potential treatment tool for controlling renal damage [13].

*T. ramosissimum* is traditionally used for the treatment of many diseases (inflammation, gastric ulcer, cancer). Its extracts markedly enhance cell proliferation either with or without mitogen (lipopolysaccharide [LPS] or lectin) stimulation and contain potent components such as flavonoids that may be potentially useful for modulating immune cell functions in physiological and pathological conditions. Moreover, *Teucrium* extract exert different protective effects against ethanol-induced ulcerogenesis [14]. Likewise, this species acts as chemopreventive and chemosensitizing agent against two uterine sarcoma cell lines, MES-SA and P-gp-overexpressing MES-SA/Dx5 cells by a slight modulation of the cell cycle and its regulators, but also through a significant induction of apoptosis [15].

## 4. Conclusions

This report affirms that phytpharmacological effects of *Teucrium* essential oil extracted at the flowering stage may be related to its derived products identified by GC/MS apparatus. *T. ramosissimum* can restore LPS-induced renal damage by inhibiting inflammation. The increase of serum biomarkers, together with glomerular and tubular alterations clearly indicate renal dysfunction.

## Acknowledgements

The authors extend their appreciation to The Ministry of Higher Education and Scientific Research of Tunisia.

## Conflict of interest

The authors declare no conflict of interest.

## Author details

Fatma Guesmi* and Ahmed Landoulsi
Faculty of Sciences of Bizerte, Laboratory of Risks Related to Environmental Stresses: Fight and Prevention, University of Carthage, Carthage, Tunisia

*Address all correspondence to: guesmif6@gmail.com

## References


[1] Maccioni A, Falconieri D, Sanna C, Porcedda S, Piras A, Maxia A. Characterization of essential oils from different taxa belonging to the genus *teucrium* in Sardinia Island, Italy. Plants. 2021;**10**:1359. DOI: 10.3390/plants10071359

[2] Napagoda M, Gerstmeier J, Butschek H, De Soyza S, Pace S, Lorenz S, et al. The anti-inflammatory and antimicrobial potential of selected ethnomedicinal plants from Sri Lanka. Molecules. 2020;**25**:1894. DOI: 10.3390/molecules25081894

[3] Jarić S, Mitrović M, Pavlović P. Ethnobotanical features of *teucrium* species. In: Stanković M, editor. Teucrium Species: Biology and Applications. Cham: Springer; 2020. DOI: 10.1007/978-3-030-52159-2_5

[4] Irshad M, Ali Subhani M, Ali S, Hussain A. Biological Importance of Essential Oils. Rijeka: Intechopen; 2019. DOI: 10.5772/intechopen.87198

[5] Fortrie G, de Geus HRH, Betjes MGH. The aftermath of acute kidney injury: A narrative review of long-term mortality and renal function. Critical Care. 2019;**23**:24. DOI: 10.1186/s13054-019-2314-z

[6] Chawla LS, Bellomo R, Bihorac A, Goldstein SL, Siew ED, Bagshaw SM, et al. Acute kidney disease and renal recovery: Consensus report of the acute disease quality initiative (ADQI) 16 workgroup. Nature Reviews Nephrology. 2017;**13**:241-257

[7] Chen Y, Jin S, Teng X, Hu Z, Zhang Z, Qiu X, et al. Hydrogen sulfide attenuates LPS-induced acute kidney injury by inhibiting inflammation and oxidative stress. Oxidative Medicine and Cellular Longevity. 2018;**2018**:10. DOI: 10.1155/2018/6717212

[8] Ben Sghaier M, Louhichi T, Ammari Y. Ethnobotanical and phytopharmacological notes on *teucrium ramosissimum* L. Research Review Biosciences. 2017;**12**(3):132

[9] Henchiri H, Bodo B, Deville A, Dubost L, Zourgui L, Raies A, et al. Sesquiterpenoids from *teucrium ramosissimum*. Phytochemistry. 2009;**70**(11-12):135-1441

[10] Farzaei MH, Zangeneh MM, Goodarzi N, Zangeneh A. Stereological assessment of nephroprotective effects of trachyspermum ammi essential oil against carbon tetrachloride-induced nephrotoxicity in mice. International Journal of Morphology. 2018;**36**(2):750-757

[11] El-Demerdash F, Dewer Y, ElMazoudy RH, Attia AA. Kidney antioxidant status, biochemical parameters and histopathological changes induced by methomyl in CD-1 mice. Experimental Toxicology and Pathology. 2013;**65**:897-901

[12] Chunzhi G, Zunfeng L, Chengwei Q, Xiangmei B, Jingui Y. Hyperin protects against LPS-induced acute kidney injury by inhibiting TLR4 and NLRP3 signaling pathways. Oncotarget. 2016;**7**(50):82602-82608

[13] Avila-Carrasco L, García-Mayorga EA, Díaz-Avila DL, Garza-Veloz I, Martinez-Fierro ML, González-Mateo GT. Potential therapeutic effects of natural plant compounds in kidney disease. Molecules. 2021;**26**:6096. DOI: 10.3390/molecules26206096

[14] Ben Sghaier M, Harizi H, Louhichi T, Krifa M, Ghedira K, Chekir-Ghedira L. Anti-inflammatory and antiulcerogenic activities of leaf extracts and sesquiterpene from *teucrium ramosissimum* (Lamiaceae). Immunopharmacology and Immunotoxicology. 2011;**33**(4):656-662. DOI: 10.3109/08923973.2011.558903

[15] Lahmar A, Mathey A, Aires V, Elgueder D, Vejux A, Khlifi R, et al. Essential oils, *pituranthos chloranthus* and *teucrium ramosissimum*, chemosensitize resistant human uterine sarcoma MES-SA/Dx5 cells to doxorubicin by inducing apoptosis and targeting P-glycoprotein. Nutrients. 1719;**2021**:13. DOI: 10.3390/ nu13051719

Chapter 10

# Essential Oil, Chemical Compositions, and Therapeutic Potential

*Slimen Selmi, Kais Rtibi, Karim Hosni and Hichem Sebai*

## Abstract

Essential oils-(EOs) are organic compounds derived from aromatic plant sources such as roots, bark, flowers, leaves and seeds. Essential oils were obtained via two different methods of extraction: steam distillation (SD) and water distillation (WD). EOs-therapy, refers to a range of traditional, alternative or complementary therapies that use essential oils from natural products and other aromatic plant compounds. The chemical components composition of EOs depends on the place of origin, climatic conditions, plant species, plant part extracted, and harvesting time. Essential oils are constituted by diversified bioactive constituents, lipophilic and volatile, and in most cases derivatives of terpene compounds and in lower occurrence phenylpropanoids. They have been long recognized for their medicinal uses: antiviral, antibacterial, insecticidal, antifungal, and antioxidant properties. This chapter provides studies on chemical composition, medicinal uses, and benefits of essential oils.

**Keywords:** essential oils, methods of extraction, bioactive chemical compositions, therapeutic potential

## 1. Introduction

Essential oils or vegetable essences are oily, volatile [1], odorous and colorless or slightly tinted products obtained by steam distillation, by expression, by incision or by enfleurage of the plant material [2]. These plant essences are widely distributed in the plant kingdom and exist only in higher plants. Indeed, they are found in appreciable quantity in approximately 2000 species divided into 60 botanical families such as for example in *Lamiaceae* (lavender, basil, mint ...), *Myrtaceae* (eucalyptus), *Lauraceae* (cinnamon and sassafras), and the *Apiaceae* (coriander, cumin, fennel, parsley..) [3]. Essential oils are found in all the organs of the plant: roots, fruits, seeds, flowers, leaves, bark, wood, etc. They are formed in specialized cells, most often, grouped in channels or in secretory pockets and they are then transported to the different parts of the plant during its growth [4].

They differ from fatty oils, by their physical properties and their composition, because they volatilize on heat and their stains on the paper are transient [5]. They are characterized by their organoleptic properties (smell, color and taste). At room temperature, they are generally liquid with a density often lower than that of water.

They are colorless or pale yellow, with a few exceptions such as the EOs of cinnamon (orange), wormwood (green) or chamomile (blue).

Their refractive index is high and most often they are endowed with rotary power. They are assigned different chemical indices (acid, ester, carbonyl number, etc.).

They are poorly soluble in water and soluble in organic solvents (ether, alcohol, hexane, pentane, etc.) [6]. They dissolve fats, iodine, sulfur, phosphorus and reduce certain salts. In addition, they oxidize and polymerize easily. To avoid this, they should be stored away from light and air.

People are beginning to use EOs widely for a variety of common conditions, and much research shows they may help relieve many disorders and their associated symptoms in some cases. The bioactive compounds in these oils may have several health benefits and actions on the human body and health. Very recent studies showed the use of these oils in many common health conditions such as anxiety, constipation, inflammation, depression and many other disruptions. The following chapter will provide more information on the novel method of extraction of EOs their chemical composition especially the main bioactive compounds, the medicinal uses, and their therapeutic benefits.

## 2. Methods of extracting essential oils

The extraction of essential oils from plant material can be carried out using many and various processes, based on ancient techniques:

- Distillation, Expression, Enfleurage or Incision or more recent: extraction under microwave or ultrasonic irradiation [6].

- Distillation remains the most popular method because it is easy to implement. **Figure 1** shows the different ways of extracting essential oils.

### 2.1 Distillation

#### *2.1.1 Hydrodistillation*

It is the simplest and most widespread technique. It involves immersing the raw material directly in water, then the whole is brought to a boil. The operation is generally carried out at atmospheric pressure. The vapors formed are condensed by a water-flow refrigeration system.

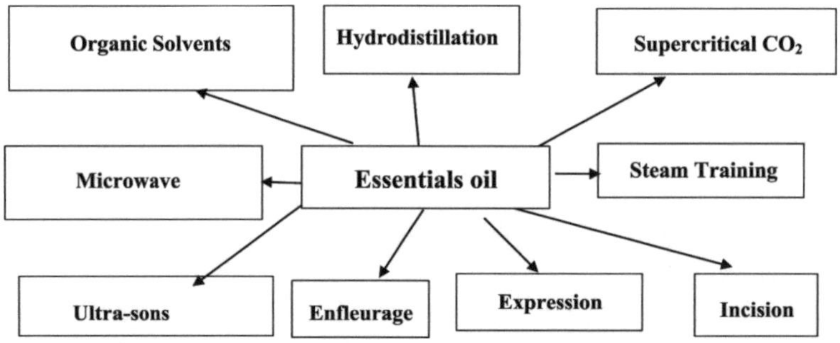


**Figure 1.**
*Methods of extracting essential oils.*

During the distillation of EOs, several phenomena are the basis of material exchanges between the solid, liquid and vapor phases, hence the influence of a large number of parameters on the quality and yield of the production of these plant essences [7].

Experiments carried out until the essence of the substrate is exhausted show that the duration of distillation is longer for the organs of woody plants than for herbaceous plants. This difference is strongly linked to the location of the production or storage systems for EOs, which are either on the surface or inside the tissues of the plant. As a result, these structures have an influence on the course of the hydrodistillation, that is to say on the successive mechanisms involved, and therefore on the duration of the extraction operation.

In the event that these structures are superficial, the outer membrane or the cuticle is quickly ruptured upon boiling, the volatile compounds are immediately evaporated. When EOs are subcutaneous, they must first diffuse through the thickness of the plant tissue before coming into contact with water or its vapor so that they can evaporate as in superficial secretions.

#### *2.1.2 Steam training*

In this type of distillation, a stream of water vapor passes through the plant which draws out the hydrophobic volatiles. After condensation, the separation takes place by settling. This method improves the quality of the EH by minimizing hydrolytic alterations.

#### *2.1.3 Distillation with organic solvents*

Some essential oils have a density close to water and the process by steam distillation cannot be used in this case. The principle consists of macerating the plant in the solvent in order to pass the odorous substances into the solvent.

##### *2.1.3.1 Petroleum based solvents*

This method uses organic solvents such as pentane, hexane, heptane, etc. It is reserved for EOs having a density close to that of water.

##### *2.1.3.2 Forane*

Forane 113 (F2CCl-CCl2F) extracts a mixture of H.E. and lipid oil at the same time, which makes it possible to double the plant.

##### *2.1.3.3 Carbon dioxide*

In liquid or supercritical carbon dioxide extraction, a stream of $CO_2$ at high pressure burst gasoline pockets [8, 9]. This method is better than hydrodistillation in terms of cost, energy saving, yield and quality of the product obtained because the carbon dioxide is colorless, odorless, non-flammable and non-toxic.

### 2.2 Microwave or ultrasonic assisted distillation

These recent techniques offer several significant advantages over conventional techniques. In fact, they require a smaller volume of solvent and a reduced heating time, which prevents the loss and degradation of volatile and heat-sensitive compounds. Thus, they lead to higher returns [6, 10].

#### *2.2.1 Microwave extraction*

Microwave extraction involves heating the extractant (water or organic solvent) in contact with the plant under microwave energy which allows for homogeneous heating. This new extraction process saves considerable time and energy [11].

#### *2.2.2 Ultrasonic extraction*

The plant material brought into contact with the solvent (water or organic solvent) is immersed in a sonication bath maintained at constant agitation [12].

### 2.3 Enfleurage

The enfleurage is a rather difficult technique. It dates from ancient Egypt and is based on the strong affinity of odorous molecules for fats. It is mainly reserved for the fragile organs that are the flowers (violet, tuberose, jasmine, ...). These are spread delicately on glass plates coated with a thin layer of grease and these plates are superimposed on wooden frames. Volatile substances diffuse and are absorbed by the fat layer. Then these facts are depleted with alcohol. This process tends to disappear because it requires a large workforce [13].

### 2.4 Expression

The expression or cold pressing is specific to the extraction of essential oils from citrus fruits: lemons, oranges, mandarins, etc. It is a fairly simple method which consists in mechanically breaking up by abrasion the gasoline pockets located at the level of the peel or pericarp of the fruit to collect its contents [14].

### 2.5 Incision

It is an infrequent operation. It is enough to split the bark of trees to collect the juice, for example the rubber of the rubber tree.

## 3. Chemical composition of EOs

The chemical composition of species is complex and can vary depending on the organism, climatic factors, the nature of the soil, cultivation practices and the method of extraction [15]. EOs are a mixture of constituents that belong to three categories of compounds: terpene, aromatic and various.

### 3.1 Terpenes

Terpenes are hydrocarbons formed by assembling two or more isoprene units. They are polymers of isoprene of the chemical formula (C5H8)n.

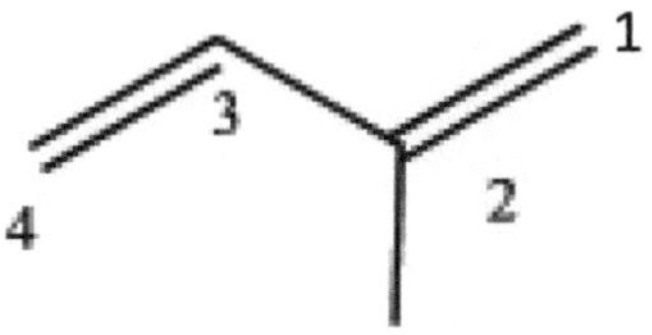


Isoprene (2methylbuta-1,3-diene)
Isoprene (2methylbuta-1,3-diene)

Depending on the number of associated units, a distinction is made between: mono- en (C10); sesqui- en (C15); di- en (C20); tri-en (C30); (C40) tetraterpenes and polyterpenes.

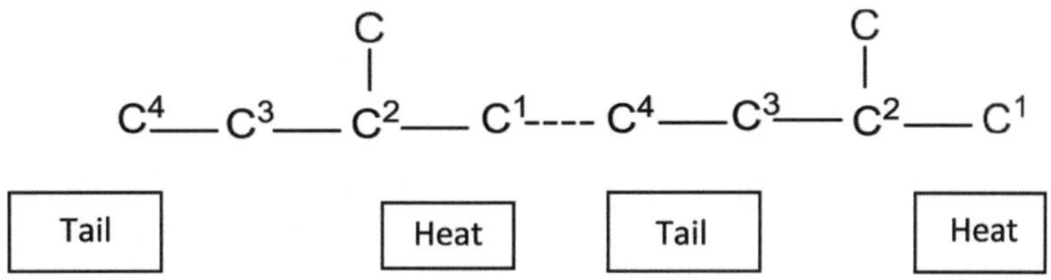


These units can bind to each other by so-called irregular bonds of the artemesyl, santolinyl, lavandulyl and chrysanthemyl type [16].

Essential oils contain particularly monoterpenes, sesquiterpenes and rarely diterpenes [17].

Terpenes have very diverse structures (acyclic, monocyclic, bicyclic, etc.) and contain most of the chemical functions of organic materials. As an indication, some structures of monoterpenes and sesquiterpenes are shown in **Figure 2**.

### 3.2 Aromatic compounds

The aromatic compounds are derived from phenylpropane (C6–C3). They are less common than terpenes. This class includes odorous compounds like vanillin, eugenol, anethole, estragole, ... (**Figure 3**). They are frequently encountered in the EOs of Apiaceae (anise, fennel, parsley, etc.) and are characteristic of those of vanilla, tarragon, basil, cloves, etc. [6]. They differ from each other by:

- The number and position of the hydroxyl and methoxy groups;
- The position of the double bond of the side chain, allylic or propenyl;
- The degree of oxidation of the aliphatic chain (alcohol, aldehyde, ketone or acid, etc).

### 3.3 Compounds of various origins

In general, low molecular weight compounds of various origins, which can be entrained during hydrodistillation, are straight or branched chain aliphatic hydrocarbons carrying different functions. As an indication, we can quote:

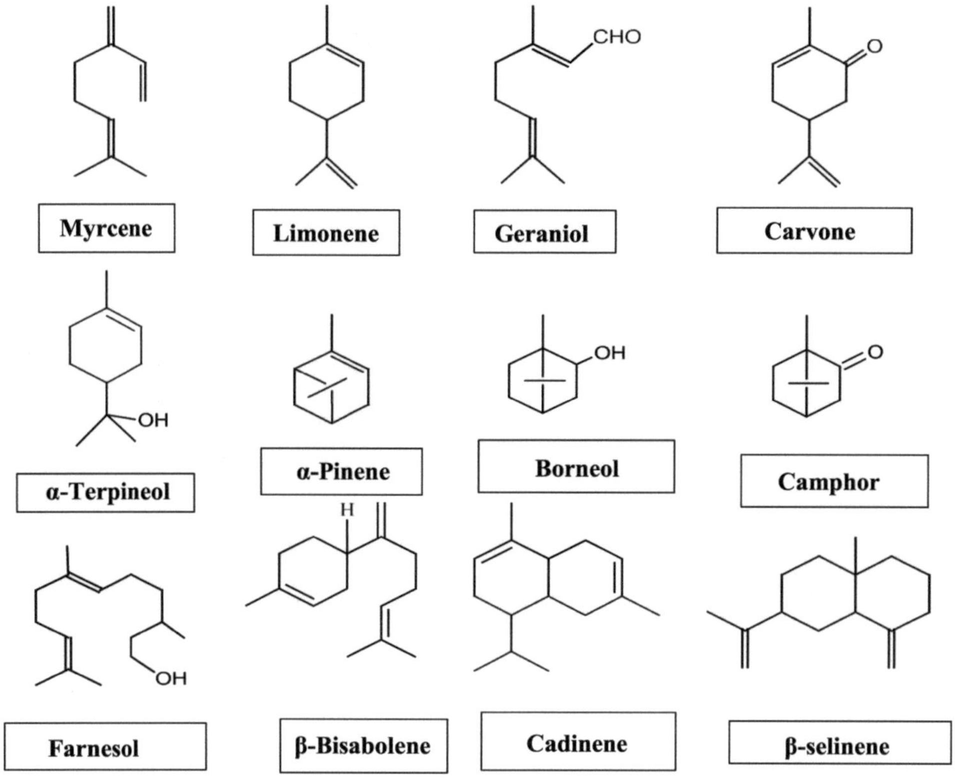


**Figure 2.**
*Examples of mono- and sesquiterpene structures.*

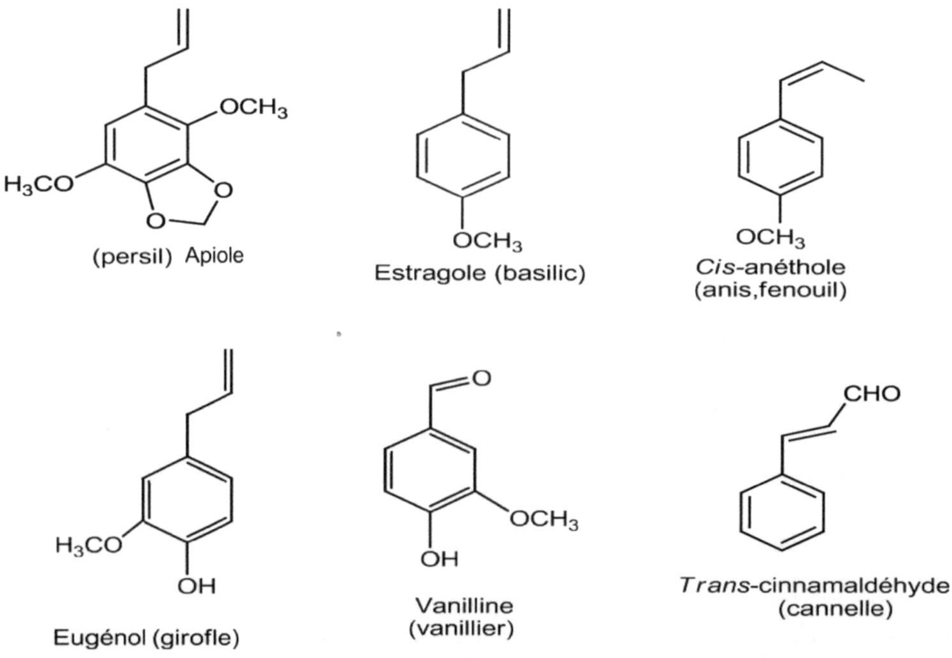


**Figure 3.**
*Examples of structures of compounds derived from phenylpropane.*

"The heptane and paraffin in chamomile oil;

- C3 and C10 acids;
- Acyclic esters found especially in fruits: butyl acetate (apple), isoamyl acetate (banana);
- Aldehydes such as octanal and decanal from Citrus;
- Alcohols such as 1-octen-3-ol in lavender oil, ...

## 4. Properties of essential oils

Essential oils have been used since ancient times for their most diverse therapeutic effects. The molecular diversity of the components they contain gives them very varied roles and biological properties [18].

In fact, monoterpene hydrocarbons have analgesic properties in percutaneous use, deworming, emmenagogue, atmospheric antiseptic, antiparasitic, etc. Sesquiterpene hydrocarbons have anti-inflammatory, calming, hypotensive effects [19].

The powers offered by the H.Es are innumerable and varied. It would be impossible to mention them all. The demonstration of their biological activity has been the subject of numerous studies [20].

### 4.1 Role of essential oils in plants

The biological role of H.Es in ecology is obvious. By their smell, they are involved in pollination. Thus, they play an attractive or repellent role with regard to predators (herbivores, insects, etc.) [7]. They can paralyze the masticatory muscles of attackers by the toxic and inappetent properties of the substances they contain [21].

They protect crops by inhibiting the multiplication of bacteria and fungi. They prevent the desiccation of the plant (loss of water) by excessive evaporation and protect the plant against light either by reduction or concentration.

Moreover, their compounds are involved in oxidation–reduction reactions, as hydrogen donors. For example, isoprene reacts rapidly with ozone and hydroxyl radicals. Also, they emit excess carbon and energy [22].

### 4.2 Biological properties

The spectrum of action of EOs is very wide, as they act against a wide range of bacteria, including those that develop resistance to antibiotics.

In addition, certain essences endowed with antifungal activity oppose the development of fungi and molds by destroying them [23]. These activities also vary from one essential oil to another and from one strain to another [24].

Essential oils act on both Gram-positive and Gram-negative bacteria. However, Gram-negative bacteria appear to be less sensitive to their action and this is directly linked to the structure of their cell wall [25] with some exceptions, such as Aeromonas hydrophila and Campylobacter jejuni, which have been described as particularly sensitive to the action of Essential oils [26]. Nevertheless, Pseudomonas aeruginosa, a Gram-negative bacterium, remains the least active vis-à-vis plant essences.

Aromatic molecules such as phenols followed by aldehydes then ketones then alcohols then ethers have the highest antibacterial coefficient. In general, the action of gasoline takes place in three distinct stages:

- Increase in permeability followed by loss of cellular constituents by attack of the essential oil on the bacterial wall;

- Blocking the production of cellular energy and the synthesis of structural components by acidification of the interior of the cell;

- Death of the bacteria by destruction of its genetic material.

### 4.3 Medicinal properties

Essential oils have many and varied medicinal properties. Most of the constituents of essential oils have antimicrobial power, hence their use as antiseptics [27]. Others have digestive or antispasmodic, sedative, healing properties, etc. These activities are mainly due to their terpene constituents.

In addition, many EOs exhibit activity against all different types of pain and are widely used to treat inflammatory joint disorders. They have the property of strengthening and reviving the individual's immune defenses [28]. It is in this sense that we could say that aromatic essences were cytophylactic (protective of living cells).

In addition, some EOs have anti-tumor activities and are adopted in the preventive treatment of certain types of cancer (Nigella, Lemon balm) [29].

## 5. Toxicity of essential oils

EOs are powerful and very active substances. They represent an inexhaustible source of natural remedies. Nevertheless, it is important to emphasize that frequent and excessive self-medication, especially with regard to the dosage as well as the mode of internal or external application by the essences is harmful. It causes more or less harmful side effects in the body (allergies, coma, epilepsy, etc.) mainly in sensitive populations (children, pregnant and breastfeeding women, elderly or allergic people) [30].

The accumulation of essences in the body by repeated intakes can lead to nausea, headaches, etc. Ingestion of more than 10 mL of essential oil is neurotoxic and epileptogenic by inhibiting the supply of oxygen to the level in brain tissue [31].

## 6. Main fields of application

Due to their various properties, EOs have become a material of considerable economic importance with a constantly growing market. Indeed, they are marketed and are of great interest in various industrial sectors such as pharmaceuticals by their antiseptic, analgesic, antispasmodic, aperitif, anti-diabetic..., in food through their antioxidant activity and flavoring effect, in perfumery and cosmetics through their odoriferous property.

### 6.1 Aromatherapy

Aromatherapy is a form of alternative medicine in which EOs are of great importance because they induce many curative effects. Thus, they are used more and more in various medical specialties such as: chiropody, acupuncture, massage-physiotherapy, osteopathy, rheumatology as well as in esthetics [4].

### 6.2 Food industry

By virtue of their antiseptic and flavoring properties, EOs are used daily in culinary preparations (garlic, bay leaf, thyme, etc.). They are also very popular in liquorice (anise drinks, kümmel) and in confectionery (candies, chocolate, etc.). Their antioxidant power allows them to preserve food by avoiding mold, preserving smen for example with thyme and rosemary [32].

### 6.3 Cosmetology and perfumery

EOs are sought after in the perfume and cosmetics industry because of their odoriferous properties. The perfume industry consumes large tonnages of essences (60%) in particular those of rose, jasmine, violet, verbena, etc. EOs are also consumed in cosmetology to perfume cosmetic products: toothpastes, shampoos, sunscreens, lipsticks, soaps [33]. Hygiene products, detergents and laundry for example, also consume a lot of EO to mask the (often unpleasant) odors of pure products.

#### *6.3.1 Pharmacy*

Essences from plants are used largely in the preparation of infusion (mint, verbena, thyme, etc.) and in the form of galenic preparations. More than 40% of medicines are based on active plant components, for example gastralgin is an anti-acid digestive which consists of EO carvi [34].

Likewise, with their flavoring properties, they help mask the unpleasant odor of drugs taken orally. Also, many drugs sold in pharmacies are based on EOs such as eye drops, creams, elixirs [34].

## 7. Conclusion

The current chapter discusses the diverse chemical bioactive compounds, the multifaceted applications of EOs in the therapeutic approach and develops its exciting potential as a novel green alternative to the toxic effects of synthetic agents. Being naturally used in food with rarely harmful actions and minimal adverse effects, EOs are exempted from toxicity aspects and accepted to be safe for preservation purposes. Also, they represent a considerable economic interest by their applications in the pharmaceutical, agro-food, cosmetological industries.

## Author details

Slimen Selmi[1*], Kais Rtibi[1], Karim Hosni[2] and Hichem Sebai[1]

1 Laboratory Functional Physiology and Bio-resources Valorisation, Higher Institute of Biotechnology of Beja, University of Jendouba, Beja, Tunisia

2 Laboratory of Natural Substances, Institut National de Recherche et d'Analyse Physico-Chimique (INRAP), Technopole Sidi Thabet, Technological Pole, Ariana, Tunisia

*Address all correspondence to: slimen.selmi@gmail.com

## References

[1] Aziz ZAA, Ahmad A, Setapar SHM, Karakucuk A, Azim MM, Lokhat D, et al. Essential oils: Extraction techniques, pharmaceutical and therapeutic potential - A review. Current Drug Metabolism. 2018;**19**(13)

[2] Bakkali F, Averbeck S, Averbeck D, Idaomar M. Biological effects of essential oils--A review. Food and Chemical Toxicology. 2008;**46**(2):446-475

[3] Nicholas A, Baijnath H. A consensus classification for the order Gentianales with additional details on the suborder Apocyninae. Botanical Review. 1994; **60**:440-482

[4] Dhifi W, Bellili S, Jazi S, Bahloul N, Mnif W. Essential oils' chemical characterization and investigation of some biological activities: A critical review. Medicines (Basel). 2016;**3**(4):25

[5] Casal S, Malheiro R, Sendas A, Oliveira BP, Pereira JA. Olive oil stability under deep-frying conditions. Food and Chemical Toxicology. 2010;**48**(10):2972-2979

[6] Siva S, Li C, Cui H, Meenatchi V, Lin L. Encapsulation of essential oil components with methyl-beta-cyclodextrin using ultrasonication: Solubility, characterization, DPPH and antibacterial assay. Ultrasonics Sonochemistry. 2020;**64**:104997

[7] Sakkas H, Papadopoulou C. Antimicrobial activity of basil, oregano, and thyme essential oils. Journal of Microbiology and Biotechnology. 2017;**27**(3):429-438

[8] Pellerin P. Perfume, Flavor. 1991;**16**(07-08):37

[9] Richard H, Loo A. La fabrication des extraits: extraction par le dioxyde de carbone, in Epices et Aromates. Journal of Supercritical Fluids. 2000;**17**:177

[10] Delazar A, Nahar L, Hamedeyazdan S, Sarker SD. Microwave-assisted extraction in natural products isolation. Methods in Molecular Biology. 2012;**864**:89-115

[11] Moradi S, Fazlali A, Hamedi H. Microwave-assisted hydro-distillation of essential oil from rosemary: Comparison with traditional distillation. Avicenna Journal of Medical Biotechnology. 2018;**10**(1):22-28

[12] Zhao Y, Fan YY, Yu WG, Wang J, Lu W, Song XQ. Ultrasound-enhanced subcritical fluid extraction of essential oil from Nymphaea alba var and its antioxidant activity. Journal of AOAC International. 2019;**102**(5):1448-1454

[13] Kanlayavattanakul M, Kitsiripaisarn S, Lourith N. J aroma profiles and preferences of Jasminum sambac L. flowers grown in Thailand. Cosmetic Science. 2013;**64**(6):483-493

[14] Samuelson R, Lobl M, Higgins S, Clarey D, Wysong A. The effects of lavender essential oil on wound healing: A review of the current evidence. Journal of Alternative and Complementary Medicine. 2020;**26**(8):680-690

[15] Najar B, Shortrede JE, Pistelli L, Buhagiar J. Chemical composition and in vitro cytotoxic screening of sixteen commercial essential oils on five Cancer cell lines. Chemistry & Biodiversity. 2020;**17**(1):e1900478

[16] Dale Poulter C, Marscle L, Hughers JM, Argyle JC. Journal of the

American Chemical Society. 1977; **99**:3823

[17] Finar IL. Organic Chemistry. Vol. II. Longman Scientific et Technical; 1994. p. 354

[18] Valdivieso-Ugarte M, Gomez-Llorente C, Plaza-Díaz J, Gil Á. Antimicrobial, antioxidant, and immunomodulatory properties of essential oils: A systematic review. Nutrients. 2019;**11**(11):2786

[19] Amor M. Les Huiles essentielles. Python-Pathos. 2006

[20] Bakkali F, Averbeck S, Averbeck D, Idaomar M. Food and Chemical Toxicology. 2008;**46**:446-475

[21] Dagli N, Dagli R, Mahmoud RS, Baroudi K. Essential oils, their therapeutic properties, and implication in dentistry: A review. Journal of International Society of Preventive and Community Dentistry. 2015;**5**(5):335-340

[22] Irshad M, Subhani MA, Ali S, Hussain A. Biological importance of essential oils. 2019. DOI: 10.5772/intechopen.87198

[23] Leyva-López N, Gutiérrez-Grijalva EP, Vazquez-Olivo G, Heredia JB. Essential oils of oregano: Biological activity beyond their antimicrobial properties. Molecules. 2017;**22**(6):989

[24] Aziz ZAA, Ahmad A, Setapar SHM, Karakucuk A, Azim MM, Lokhat D, et al. Essential oils: Extraction techniques, pharmaceutical and therapeutic potential - A review. Current Drug Metabolism. 2018;**19**(13):1100-1110

[25] Solórzano-Santos F, Miranda-Novales MG. Essential oils from aromatic herbs as antimicrobial agents. Current Opinion in Biotechnology. 2012;**23**(2):136-141

[26] Ozogul Y, Kuley E, Ucar Y, Ozogul F. Antimicrobial impacts of essential oils on food borne-pathogens. Recent Patents on Food, Nutrition & Agriculture. 2015;**7**(1):53-61

[27] Kalemba D, Kunicka A. Antibacterial and antifungal properties of essential oils. Current Medicinal Chemistry. 2003;**10**(10):813-829

[28] AK, Pandey VV, Beg S, Rawat JM, Singh A. Biological, medicinal and toxicological significance of Eucalyptus leaf essential oil: A review. Journal of the Science of Food and Agriculture. 2018;**98**(3):833-848

[29] Ramsey JT, Shropshire BC, Nagy TR, Chambers KD, Li Y, Korach KS. Essential oils and health. The Yale Journal of Biology and Medicine. 2020;**93**(2):291-305

[30] Vigan M. Essential oils: Renewal of interest and toxicity. European Journal of Dermatology. 2010;**20**(6):685-692

[31] López V, Nielsen B, Solas M, Ramírez MJ, Jäger AK. Exploring pharmacological mechanisms of lavender (Lavandula angustifolia) essential oil on central nervous system targets. Frontiers in Pharmacology. 2017;**8**:280

[32] Han JH, Patel D, Kim JE, Min SC. Retardation of listeria monocytogenes growth in mozzarella cheese using antimicrobial sachets containing rosemary oil and thyme oil. Journal of Food Science. 2014;**79**(11):E2272-E2278

[33] Cavanagh HM, Wilkinson JM. Biological activities of lavender essential oil. Phytotherapy Research. 2002;**16**(4):301-308

[34] da Silva JK, da Trindade R, Alves NS, Figueiredo PL, Maia JGS, Setzer WN. Essential oils from Neotropical Piper species and their biological activities. International Journal of Molecular Sciences. 2017;**18**(12):2571

Chapter 11

# Essential Oils High in 1,8-Cineole of Mediterranean Flavoring Plants: Health Benefits

*Sílvia Macedo Arantes, Ana Teresa Caldeira and Maria Rosário Martins*

## Abstract

Aromatic flavoring plants are important ingredients of the Mediterranean diet, one of the healthiest and most sustainable dietary forms, often associated with greater longevity as well as contributing to the reduction of some chronic pathologies with high mortality and morbidity. Their essential oils (EOs) are increasingly used as therapeutic agents and food supplements, due to their antioxidants, anti-inflammatory or anti-tumoral properties. The Health benefits of essential oils are closely related with their chemical constituents. The 1,8-cineole, a naturally cyclic oxygenated monoterpene, has been attributed several biological properties such as antioxidants, anti-inflammatory or antitumoral. Nevertheless, the EO properties are attributed not only to their main components but also to the synergistic effect of minor components. This review chapter focused on the chemical composition and antioxidant and anti-inflammatory potential of EOs of flavoring Lamiaceae plants, with high content in 1,8-cineole, including chemotypes of genera *Lavandula*, *Calamintha*, *Rosmarinus,* and *Thymus*, often used in the Mediterranean diet.

**Keywords:** natural products, 1,8-cineole, antioxidants, anti-inflammatory properties

## 1. Introduction

Aromatic plants are increasingly used as therapeutic agents and as food supplements, along with industrial synthesis products. The World Health Organization (WHO) estimates that more than 80% of the world population uses products based on plant extracts and/or their active components for various purposes, including health care and phytotherapy [1–3].

Essential oils (EOs) are volatile compounds, products of secondary metabolic processes of aromatic plants and despite being practically insoluble in water, can be carried away by water vapor. They are largely obtained by water distillation or using steam distillation, from different parts of the plant, including the whole plant or just the wood, roots, leaves or flowers [4, 5]. Other processes to obtain oils from plants include expression, solvent extraction, $CO_2$ extraction, maceration, cold pressure extraction [6]. Indeed, the species, the plant geographical conditions, and the part of the plant used as well as the extraction method used will be determinants for the EOs chemical profile [7–11]. Otherwise, in the distillation process, thermal degradation of sensitive compounds, the photo-oxidation of light-sensitive compounds or the hydrolysis of esterified compounds are factors that can affect the chemical profile of EOs [7–12].

Aromatic, spice and medicinal plants are part of the Mediterranean diet, recognized by the WHO as a healthy and health-promoting type of diet [1, 2]. They also represent a growing interest in the food industry and are often used in alternative or complementary therapies in conventional medicine [13, 14]. Due to their effectiveness and, mainly, due to the lower number of adverse effects, when compared to synthetic drugs, the use of aromatic plants as functional foods as well their EOs may be an important role in the prevention of pathologies with high mortality and morbidity, such atherosclerosis, neurodegenerative diseases, diabetes, several infections, chronic inflammatory diseases, cancer and autoimmune diseases [15–20]. Some flavoring plants used in the Mediterranean diet are frequently used in Alentejo (South of Portugal) as food additives or flavors. Most of them belong to the Lamiaceae family and include *Lavandula* spp., *Calamintha* spp., *Rosmarinus* spp. and *Thymus* spp., in which EOs show chemical polymorphism with high content in 1,8-cineole and that are recognized for their antioxidant and anti-inflammatory potential [21–36].

The genus *Lavandula* comprises about 32 species, commonly known as lavender, such as *L. angustifolia and L. latifolia*, as well as Mediterranean *L. stoechas*, *L. pedunculata* and *L. viridis,* often found in the southern region of Portugal. Due to their great diversity, some species have a difficult taxonomic classification due to their hybridization capacity and morphological diversity, being important to characterize them by the composition of their OEs due to their great economic importance. *Lavandula* EOs are generally produced by distillation, either from the flower spike or from the leaves [37–39]. *Lavandula. stoechas* L. subsp. *luisieri* (Rozeira) Rozeira, *Lavandula pedunculata* (Mill.) Cav. and *Lavandula. viridis* L'Hér, are endemic to the Iberian Peninsula and wild growth in some regions the southern region of Portugal [40]. *L. luisieri* and *L. pedunculata* can be distinguished mainly by the shapes of the bracts the length of the stalks of the ears as well as by the diversity of their EOs components [21, 41, 42]. *L. viridis*, known as green lavender or white lavender, and their EOs showed antioxidant and anti-inflammatory activities, depending on the plant polymorphism and the EO chemotype [4, 21, 43–46].

The genus *Calamintha* consists of eight native species belonging to the Lamiaceae family, of which six species are extremely polymorphic [47]. *Calamintha nepeta* subsp. *nepeta* (syn. *Clinopodium nepeta* (L.) Kuntze) commonly known as neveda or calamint, is a perennial and quite aromatic plant and is widely distributed in the Mediterranean area [4, 40]. It is well known as a spice used in food flavoring and as an antiseptic and diuretic stimulant [48]. It is traditionally used in medicine as an antitussive and expectorant; it also has spasmolytic and anti-flatulence properties [49, 50]. Some studies report that its EO has antifungal and antibacterial activity, due to the high content of terpene derivatives [48]. It has also been reported that the EOs of this species have antioxidant, antimicrobial, anti-inflammatory and sedative properties [51–54].

The genus *Rosmarinus* (Lamiaceae) that grows wild in the western Mediterranean region is composed of three different species: *Rosmarinus officinalis*, *Rosmarinus eryocalix* and *Rosmarinus tomentosus* [55, 56]. *Rosmarinus officinalis* L. (Syn: *Salvia rosmarinus* Schleid and *Rosmarinus angustifolius* Mill) commonly known as rosemary is widely used worldwide and is indigenous from the Mediterranean region, spontaneous in heaths, thickets and pine forests in the Center and South of the Continent [57–59]. Hepatoprotective, antioxidant and antimicrobial effects are attributed to this medicinal plant, as well as action in rheumatic diseases and digestive problems [59, 60].

The genus *Thymus* (Lamiaceae), also widely distributed in the Iberian Peninsula, is a taxonomically complex group of aromatic plants, traditionally used for medicinal purposes and contains about 214 species throughout the world [61–67]. *Thymus mastichina* (L.) is an endemic species of the Iberian Peninsula and it is found from north to south of Portugal (generally at the interior north and at the south) [4]. Commonly known as mastic thyme or Spanish marjoram, it is characterized by simple and opposite leaves and zygomorphic and bilabial flowers. *Thymus mastichina* L. subsp. *mastichina* has great ecological plasticity and is usually present in clearings of xerophytic bushes, roadsides, and slopes, abandoned fields, pine forests, cork oaks, stony areas and rocky outcrops. They prefer removed substrates, generally siliceous, quite sandy, and also schist and limestone substrates [68]. It is widely used for its medicinal properties, including antiseptic, digestive, antirheumatic, antispasmodic, expectorant and antitussive effects, and is also used as a flavoring plant in perfume and cosmetics industry [4, 68, 69]. *Thymus capitellatus* Hoffmanns & Link, commonly known as "*wild-thyme*", is an aromatic species endemic to the southern of Portugal, growing in sandy substrates of the Tagus and Sado basins (Extremadura, Ribatejo and Alentejo provinces of Iberian Peninsula) [70].

## 2. *Inter*- and *intra*-specific differences in EOs compositions

EOs are an important source of bioactive compounds with application in phytotherapy and traditional medicine. They are volatile complex compounds characterized by a strong odor and are rich in terpene compounds, namely monoterpenes (C10) and sesquiterpenes (C15), although diterpenes (C20) may also be present, as well as a variety of low molecular weight aliphatic hydrocarbons, acids, alcohols, aldehydes, acyclic esters or lactones and, exceptionally, compounds containing nitrogen (N) and sculpture (S), coumarins and phenylpropanoid homologs [6, 71, 72].

Among the different terpenes present in EOs, 1,8-cineole or eucalyptol (1,3,3-trimethyl-2-oxabicyclo [2.2.2] octane) in cyclic monoterpene oxide with a strong odor well known as the major constituent (>70%) of diverse eucalyptus species [73–78]. Some studies have demonstrated the high pharmacological potential of 1,8-cineole, namely as an antioxidant and anti-inflammatory compound [73–81]. With no negative effects in animal experiments, 1,8-cineole is considered safe when administered at normal doses (very high value of $LD_{50}$ in rats - between 1.5 and 2.5 g/kg) [82]. Nevertheless, EOs of some Lamiaceae flavoring plants showed high content in 1,8-cineole, such as some lavenders (*Lavandula* ssp.) [21–23, 25, 42, 43, 45, 83–90], rosemary (*Rosmarinus officinalis*) [29, 30, 58, 91], *Calamintha nepeta* [26, 27, 50, 92, 93] and some *thymes* (*Thymus mastichina* [26, 27, 36, 94, 95] and *Thymus capitellatus* [70, 96]).

Due to their natural function, the chemical composition of EOs is determined not only by the genus, species, and subspecies of an aromatic plant but also by external factors such as geographic location, environmental conditions of the region, cultivation conditions, season and time of harvest [7–10]. In addition, it is also necessary to consider some procedures, such as techniques of plant collection or post-harvest conservation, part of the plant used, and the EOs extraction method also affects the chemical composition of EOs [7–10].

**Table 1** presents some of the main components of EOs of aromatic plants from the Mediterranean region of the genera *Lavandula*, *Calamintha*, *Rosmarinus* and *Thymus,* described in the bibliography. EOs of these flavoring plants present polymorphisms and, consequently, it is possible to find at least two or three chemotypes.

Some studies report that *L. luisieri* EOs has a unique composition in the Plantae kingdom, containing irregular cyclopentene monoterpenes derived from necrodane, such as α-necrodol, α-necrodyl acetate; in addition to, 1,8-cineole, lavandulyl acetate, α-pinene, linalool, camphor and fenchone [83, 86, 87, 97]. However, the chemical compositions of the EOs of *L. stoechas* subsp. *stoechas* and *L. stoechas* subsp. *luisieri* are quite distinct [85, 87], so this classification has been controversial studies carried out with the EOs of *L. luisieri* from southern Portugal, revealed a chemical profile rich in oxygenated monoterpenes (>50%) and hydrocarbons sesquiterpenes (5–11%), of which 1,8-cineole (18.8%), *trans*-α-necrodyl acetate (16.2%), lavandol (11.7%), *trans*-α-necrodol (10.6%), and *β*-caryophyllene (6.0%) [23]. Recently, a study with EOs of *L. luisieri* and *L. pedunculata* from Portugal reported EOs of *L. luisieri* with 1,8-cineole (6–34%); fenchone (0–18%) and α-Necrodyl acetate (3–17%); and *L. pedunculata* with major compounds 1,8-Cineole (12–34%); fenchone (6–50%) and camphor (10–34%) (three chemotypes: 1,8-cineole, fenchone and camphor) [88]. In studies carried out by Garcia-Vallejo, *et al.* [98] and Lavoine-Hanneguelle and Casabianca [87] with *L. luisieri* from Spain, the EOs presented as main compounds 1,8-cineole, lavandulol, lavandulyl acetate, linalool and their acetates, also present in other species of the genus *Lavandula*, in addition to some necrodane compounds. According to Miguel et al. [21] the populations of Spain showed high levels of 1,8-cineole, fenchone, camphor and necrodane derivates and, in the south of Portugal, the majority compound it was always the 1,8-cineole [84, 86].

EO of *L. pedunculata* showed high content of oxygenated monoterpenes, although with quantitative differences regarding the percentages of the various compounds [21, 90]. Studies carried out with *L. pedunculata* from central Portugal [90] indicate that its EO consists mainly of oxygenated monoterpenes (69–89%) and hydrocarbons monoterpenes (4.25–22.5%), with fenchone as the main constituents (1.3–59.7%), 1.8-cineole (2.4–55.5%) and camphor (3.6–48.0%). The EOs of *L. pedunculata* from central Portugal have been categorized into three chemotypes: 1,8-cineole, 1,8-cineole/camphor and fenchone [35], while the EO of *L. pedunculata* from the Algarve (Portugal) expresses the camphor/camphene chemotype [25].

Analysis of the EO of *L. viridis* revealed a chemical composition with a predominance of terpenoids, namely oxygenated monoterpenes (>50%) and hydrocarbons monoterpenes (>20%) and sesquiterpenes (<5%), presenting as majority: 1,8-cineole, camphor, α-pinene and linalool, the leaves of which are used, dried, in medical applications in Madeira, Portugal [43–46]. Some authors relate that the EO of the aerial part (leaf and flower) of this species contains mostly oxygenated monoterpenes (>50%), hydrocarbons monoterpenes (>20%) and sesquiterpenes (<5%), presenting as major compounds: 1,8-cineole (22–42%), camphor (2.9–31.5%), α-pinene (0.3–14.4%) and linalool (0.2–7.8%) [21, 43–45].

Depending on the variety and region, *Calamintha* EOs showed as major components carvacrol (45–65%), β-pinene, geraniol, α-caryophyllene and pulegone [48–50]. Previous studies with EOs oils from *C. nepeta* indicate the presence of a remarkable chemical polymorphism, suggesting the existence of two chemotypes: one characterized by the predominance of pulegone and menthone, menthol and/or its isomers, piperitenone, piperitone and their oxides [93, 99–105], and the other type characterized by the predominance of piperitenone oxide and/or piperitone oxide [93, 106]. Marongiu

| EO | Country | Part used | Major volatile compounds | Ref. |
|---|---|---|---|---|
| *Lavandula luisieri* | Portugal (Algarve) | Flowering aerial parts | 1,8-Cineole (26–34%); α-Necrodyl acetate (11–18%) | [21] |
| | Portugal (Penamacor) | Flowers | 1,8-Cineole (3–4%); Camphor (8–21%); Linalool (1.4–3%); α-Necrodil acetate (2–20%) | [22] |
| | Portugal (Penamacor) | Leaves | 1,8-Cineole (13.9–16.4%); Camphor (1–3%); Linalool (1–2%); α-Necrodil acetate (8–19%) | [22] |
| | Portugal (Alentejo) | Flowering aerial parts | 1,8-Cineole (19%); α-Necrodyl acetate (16%); Lavandulol (12%); α-Necrodol (11%); *β*-Caryophyllene (6%) | [23] |
| | Portugal (Piódão region) | Flowering aerial parts | 1,8-Cineole (6.4%); α-Necrodyl acetate (17%) | [83] |
| | Portugal (Algarve) | Flowering aerial parts | 1,8-Cineole (34%); Fenchone (18%); α-Necrodyl acetate (3%) | [83] |
| | Spain (Toledo; Sevilha) | Flowering aerial parts | 1,8-Cineole (0.4–21%); Fenchone (1.4–22%); Camphor (2–54%) | [84] |
| | Spain (Sevilha) | Flowering aerial parts | α-Necrodol, α-Necrodyl acetate and 1,8-Cineole, (>50%) | [85] |
| | Spain | Leaves flowers | Camphor (81%); 1,8-Cineole (77%) Camphor (88%); 1,8-Cineole (85%) | [86] |
| | Spain (Sevilha) | Flowering aerial parts | 1,8-Cineole (16%); α-Necrodyl acetate (23%) | [87] |
| | Portugal | Flowering parts | 1,8-cineole (6–34%); fenchone (0–18%); α-Necrodyl acetate (3–17%); | [88] |
| *Lavandula pedunculata* | Portugal (Algarve) | Flowering aerial parts | Fenchone (42–44%); Camphor (35–36%) | [21] |
| | Portugal (Algarve) | Flowering aerial parts | Camphor (41%); Fenchone (38%) | [25] |
| | Portugal (Santarém) | Flowering aerial parts | Fenchone (62–70%); 1,8-Cineole (6–28%) | [89] |
| | Portugal (Trás-os-Montes) | Flowering aerial parts | 1,8-Cineole (24%); Camphor (32.4%) | [42] |
| | Portugal (Coimbra) | Flowering aerial parts | Fenchone (49%); α-Pinene (5%); α-Cadinol (4%) | [42] |
| | Portugal (North and Center) | Flowering aerial parts | 1,8-Cineole (2.4–56%); Fenchone (1.3–60%); Camphor (4–48%) | [90] |
| | Portugal | Flowering parts | 1,8-Cineole (12–34%); fenchone (6–50%); Camphor (10–34%); | [88] |

| EO | Country | Part used | Major volatile compounds | Ref. |
|---|---|---|---|---|
| *Lavandula viridis* | Portugal (Algarve) | Flowering aerial parts | 1,8-Cineole (35%); Camphor (13%); α-Pinene (9%) | [45] |
| | Portugal (Algarve) | Flowering aerial parts | 1,8-Cineole (33%); Camphor (20%) | [21] |
| | Portugal (Algarve) | Flowering aerial parts; | 1,8-Cineole (18–25%); Camphor (9–12%); Borneol (4–5%); α- terpineol (1–4%) | [43] |
| *Calamintha nepeta* | Portugal (Algarve) | Aerial parts | 1,8-Cineole (30%); Isopulegone (36%) | [26] |
| | Portugal (Alentejo) | Flowering aerial parts | 1,8-Cineole (28%); Menthone (22%); Menthol (16.3%) | [27] |
| | Italy (Basilicata region) | Flowering aerial parts | Pulegone (45%); Menthone (16%); Piperitenone (13%); Piperitone (6%) | [92] |
| | Italy (Sardinia Island) | Flowering aerial parts | Pulegone (40–64%); Piperitenone (6–8%); Piperitenone oxide (2.5–19%) | [50] |
| | Portugal | Flowering aerial parts | Isomenthone (36–51%); 1,8-Cineole (21%); *trans*-Isopulegone (6–8%) | [50] |
| | Serbia | Flowering aerial parts | Pulegone (76%) | [93] |
| *Rosmarinus oficinallis* | Si Chuan Province, China | Aerial parts | 1,8-Cineole (27%); α-Pinene (19%); Camphor (14%); Camphene (12%); β-Pinene (7%) | [29] |
| | Iran | Aerial parts | α-Pinene (15%); 1,8-Cineole (7%); Linalool (15%), | [30] |
| | Portugal (Vila Real) | Aerial parts | 1,8-Cineole (23–67%) | [58] |
| | Algeria | Leaves and stems | 1,8-Cineole (leaves: 54%,stem: 30%) | [91] |
| *Thymus mastichina* | Portugal (Algarve) | Flowering aerial parts | 1,8-Cineole (47–61%) | [94] |
| | Portugal (Algarve) | Aerial parts | 1,8-Cineole (41%) | [26] |
| | Italy | Micropropagated plantlets | 1,8-Cineole (58%); Linalool (25%) | [95] |
| | Spain (Murcia) | Flowering aerial parts | 1,8-Cineole (39–74%); Linalool (2.2–43%) | [36] |
| | Portugal (Alentejo) | Flowering aerial parts | 1,8-Cineole (71%) | [27] |

| EO | Country | Part used | Major volatile compounds | Ref. |
|---|---|---|---|---|
| *Thymus capitellatus* | Portugal (Ribatejo) | Aerial parts (flowering) | 1,8-Cineole (59%); Borneol (10%) | [96] |
| | Portugal (Porto Alto) | Aerial parts | 1,8-Cineole (48%) | [70] |
| | Portugal (samora Correia) | Aerial parts | 1,8-Cineole (29%); Borneol (29%) | [70] |
| | Portugal (Poceirão) | Aerial parts | 1,8-Cineole (28%); Linalyl acetate (20%);<br>Linalool (17%) | [70] |

**Table 1.**
*Main components present in some EOs of flavoring plants.*

*et al.* [50], in a study comparing the chemical profile carried out with Portuguese and Italian *C. nepeta*, reported that the EO of Portuguese *C. nepeta* presented as major components isomenthone, 1,8-cineole and isopulegone, while the Italian EO presented pulegone as a major component. Studies carried out with *C. nepeta* from southern Portugal reported that this EO had 1,8-cineole, isopulegol and isopulegone as major components [26]. Studies carried out with *C. nepeta* EOs from Alentejo (Portugal) have shown a peculiar chemical profile predominantly composed of isomenthone (35.8–51.3%), 1,8-cineole (21.1–21.4%) and trans-isopulegone (7.8–6.0%) [27, 28, 107–109].

*Rosmarinus* EOs present α-pinene (up to 30%), β-pinene, camphene, limonene, myrcene, β-caryophyllene, cineole (15–30%), camphor (15–25%) as major compounds [4, 34]. The EO of *R. officinalis* presents a chemical polymorphism. In studies carried out by Ribeiro et al. [58], Wang et al. [29] and Boukhobza et al. [91] with the EO of *R. officinalis* from Portugal, China and Algeria, respectively, the EO presented as the majority compound 1,8-cineole. The EO of *R. officinalis* from Iran presented, as major components, α-pinene and linalool, with 14% each [30].

*Thymus vulgaris*, a flowering plant from genus *Thymus* and originating in southern Europe, is characterized by chemical component polymorphism according to the main volatile, with known six EOs chemotypes: geraniol, linalool, α-terpineol, tujanol-4, thymol and carvacrol [110]. *T. mastichina* is a related species to the *Thymus* genus but it presents a chemical polymorphism, with 1,8-cineole, limonene and β-terpinol as main constituents [4]. According to Salgueiro et al. [111], the EOs of some thyme species were characterized by a high content of 1.8-cineole and variable content of linalool, which varies with the geographical origin. Another study with EOs from *T. mastichina* reported 1,8-cineole (64%) as a major component, followed by α-terpineol (6%) and β-pinene (5%) [4, 112]. Moreover, Portuguese thyme from the *T. mastichina* section has also 1,8-cineole as the main constituent (often higher than 60%) [111]. Studies with a related species, *T. capitellatus* reveal the existence of polymorphisms in their EOs, reporting three chemotypes: the 1,8-cineole; the 1,8-cineole/borneol and the 1,8-cineole/linalyl acetate/linalool chemotypes [70].

## 3. Health benefits of EOs

Plants and their extracts have been used by mankind since the beginning of history and their secondary metabolites have traditionally played an important role in human health and well-being [71], increasingly important in therapeutics, due to their efficacy and, above all, due to the lower number of adverse effects when compared to synthetic drugs.

**Table 2** reports some pharmacological activities (antioxidant analgesic, anti-inflammatory and cholinesterase inhibition) of EOs of some *Lavandula* spp., *Calamintha nepeta*, *Rosmarinus officinallis* and *Thymus mastichina* chemotypes high in 1,8-cineole. Depending on their chemical composition, either major and minor constituents, studies report antioxidant, antitumor, analgesic and anti-inflammatory, sedative or antispasmodic effects of these EOs [72, 113].

### 3.1 Antioxidant activity of EOs

The adverse effects of oxidative stress on human health have become a serious issue. This results from the imbalance between oxidant and antioxidant molecules, which can induce cellular damage by free radicals and promote the development of many current disease conditions, including inflammation, autoimmune diseases, cataracts, cancer, Parkinson's disease, arteriosclerosis and aging [1, 114]. Reactive oxygen species (ROS) are constantly generated and play important roles in a variety of normal biochemical functions as well as irregular or pathological processes. Furthermore, ROS can be produced by a family of mitochondrial membrane-bound enzymes, such as NAD(P)H oxidases, which appear to affect cell proliferation and apoptosis [115].

A broad definition of an antioxidant is "any substance which, present in low concentrations compared to that of the oxidizable substrate, effectively delays or inhibits the oxidation of that substrate". EOs are important antioxidants able to prevent or minimize the development of degenerative diseases, including cardiovascular diseases, cancer, neurodegenerative and inflammatory diseases [1].

Many of the medicinal plants belonging to the Lamiaceae family have antioxidant potential. Studies carried out with medicinal plants suggest that their antioxidant activity is due to the redox reactions of phenolic compounds, which allow them to act as reducing agents, donating hydrogen atoms and capturing singlet oxygen [116].

Some studies carried out with species of the genus *Lavandula* suggests that EOs from the aerial parts of these plants have antioxidant activity in protecting the lipid substrate and capturing free radicals, depending on their chemical constituents [21, 22, 37, 117, 118].

EOs of *C. nepeta* from Portugal have *in vitro* antioxidant capacity either to capture free radicals and to reduce $Fe^{3+}$ or inhibit lipid oxidation [26–28]. Also, the EO of *R.*

| EOs | Biological activities | Ref. |
|---|---|---|
| *Lavandula luisieri* | Antioxidant and analgesic or anti-inflammatory potential | [21–24, 88] |
| *Lavandula pedunculata* | Antioxidant activity and cholinesterases inhibition | [21, 25, 88] |
| *Lavandula viridis* | Antioxidant activity | [21] |
| *Calamintha nepeta* | Antioxidant and Antitumoral potential | [26–28] |
| *Rosmarinus oficinallis* | Antioxidant activity and acetylcholinesterase inhibition; Antiproliferative activities | [29–35] |
| *Thymus mastichina* | Antioxidant potential and Acetylcholinesterase inhibition, Lipoxygenase inhibition and anti-tumoral activities | [26, 27, 36] |

**Table 2.**
*Biological properties of EOs with high content in 1,8-cineole.*

*officinallis* with high content of 1,8-cineole reported that it could inhibit lipid peroxidation and capture free radicals [29, 31, 32, 34]. Studies carried out with *T. mastichina* EOs from Spain and Portugal showed that their EOs have shown low antioxidant activity by the DPPH radical method [26, 36, 119].

EOs are an important source of potentially useful antioxidants to prevent oxidative stress and promote human health [120]. According to the literature, the antioxidant activity of EOs is related to their high content of monoterpenes, namely limonene, 1,8-cineole, γ-terpinene, α-terpinene, linalool, 4-terpineol [60, 121, 122]. Additionally, the synergistic potential of minority constituents is often proposed to explain the differences between estimated and observed values for antioxidant capacities [123, 124]. Antioxidant properties of EOs also suggest their potential as anti-inflammatory agents, since the capture and elimination of free radicals is one of the mechanisms involved in the prevention of inflammation [125, 126]. Additionally, due to their high activity in protecting the lipid substrate, EOs have the potential to prevent neurodegenerative and cancerous diseases [35, 127].

### 3.2 Anti-inflammatory activity of EOs

The inflammatory response is one of the most important defense mechanisms of the body, responsible for removing and neutralizing invading microorganisms and/or repairing tissues, involving, in its processes, immune cells of the hematopoietic system, such as macrophages. Cyclooxygenases (COX) play an important role in mediating the body's inflammatory response [128–130]. Cytokines released in the anti-inflammatory processes (IL-1, IL-2, IL-6, IL-8 and TNF or tumor necrosis factor) are also associated with other body responses, including immune and anti-inflammatory responses or anti-tumoral and apoptosis processes [131–134].

EOs of several plants promote anti-inflammatory activity due to the presence of bioactive compounds such as oxygen monoterpenes that mediate the capture of free radicals generated by neutrophils and macrophages as well as for their ability to inhibit the cyclooxygenase pathway, having an important role in the regulation of inflammatory mediators [126, 135]. There is evidence that the association of chronic inflammation and oxidative stress with the aging process, indicating a subclinical chronic response. Additionally, the presence of reactive species in inflammatory processes are present in the etiology of several pathologies, including those resulting from metabolic disorders, demonstrating the central role of the reciprocal interaction between oxidative stress and inflammation [136–141].

The anti-inflammatory activity of EOs can be attributed not only to their antioxidant properties but also to interactions with signaling cascades involving cytokines and transcriptional regulatory factors and in the expression of pro-inflammatory genes [126, 142]. Some studies carried out in animal models with terpenes present in EOs, such as linalool, limonene, myrcene, 1,8-cineole, demonstrated that these compounds showed analgesic activity [143–152]. The analgesic and anti-inflammatory potential of OEs are preferentially attributed to the high content of terpene compounds [143, 145, 146, 153–156] as well as to the synergetic effect of minor components that can influence the pharmacokinetics and bioavailability of compounds with pharmacological action [157].

The anti-inflammatory effects observed for the EOs of *Lavandula* species can be attributed to their monoterpene content, namely 1,8-cineole, fenchone, linalool [143–147, 152]. For example, a study with EO from the leaves of *L. angustifolia*, high in 1,8-cineole (65%), borneol (12%) and camphor (10%) reported EO anti-inflammatory activity of 48% at a dose of 200 mg/kg [38]. In another case, Cardia et al. [158] demonstrated that the EO of *L. angustifolia* has anti-inflammatory activity by the method of paw edema induced by carrageenan, being able to inhibit, at a dose of 100 mg/kg, the inflammation in 54%, 56% or 45% after 30, 60 or 120 min, respectively. Recently, Zuzarte et al. [88] evaluated the anti-inflammatory activity of EOs of *L. luisieri* (high content in 1,8-cineole and fenchone and low quantities of necrodane derivatives) and three different EOs chemotypes of *L. pedunculata* (1,8-cineole, fenchone and camphor chemotypes) from Portugal and reported that EOs of *L. luisirei,* rich in 1,8-cineole and fenchone, was the EO with the highest anti-inflammatory potential and also was more active than its major compounds when assessed alone or in combination, confirming the synergetic effect of minor components.

Regarding the species of the genus *Thymus*, the most studied species is *T. vulgaris*, being widely recognized for the potential of its EOs and their major components as anti-inflammatory agents [159–161]. On the other hand, the literature reports that the monoterpenes 1,8-cineole, anethole and fenchone, major components present in the EOs of *C. nepeta* and *T. mastichina* may be also responsible for the anti-inflammatory effect [77, 143, 144, 147–152, 162].

## 4. Conclusions

EOs are increasingly used as therapeutic agents, cosmetics and food additives, along with industrial synthesis products, with application in phytotherapy. However, several factors can affect the biological properties of OEs, such as genera and species, time and region of harvest, extraction method, as well as the polymorphisms of each species. Correlation study between the biological properties of EOs and their chemical composition allows to evaluate its phytopharmaceutical potential and, together with traditional knowledge and practices, scientifically validate it, to allow an adequate, effective and safe use.

The biological activities of EOs are often related with its high content of some monoterpenes, such as 1,8-cineole which is an oxygenated monoterpene frequently found as one of the major components in the EOs of some plants of genera *Lavandula*, *Calamintha*, *Rosmarinus* and *Thymus*, autochthonous to the Mediterranean region. Several biological properties of EO of these plants have been attributed, including antioxidant (capacity to capture of free radicals or ability to protect the lipid substrate) and anti-inflammatory properties. These EOs properties are often attributed to their major components, however, the synergistic potential of minor constituents is often proposed to explain the differences between estimated and observed values.

## Acknowledgements

This work was supported by the UIDB/04449/2020 and UIDP/04449/2020 projects, funded by Fundação para a Ciência e Tecnologia (FCT).

## Conflict of interest

The authors declare no conflict of interest.

## Author details

Sílvia Macedo Arantes[1], Ana Teresa Caldeira[1,2,3] and Maria Rosário Martins[1,4*]

1 Hercules Laboratory, Institute of Research and Advanced Training (IIFA), University of Evora, Evora, Portugal

2 Department of Chemistry and Biochemistry, School of Science and Technology, University of Evora, Evora, Portugal

3 City U Macau Chair in Sustainable Heritage, Institute of Research and Advanced Training (IIFA), University of Evora, Evora, Portugal

4 Department of Medical and Health Sciences, School of Health and Human Development, University of Evora, Evora, Portugal

*Address all correspondence to: mrm@uevora.pt

## References


[1] Krishnaiah D, Sarbatly R, Nithyanandam R. A review of the antioxidant potential of medicinal plant species. Food and Bioproducts Processing. 2011;**89**(C3):217-233

[2] Ssegawa P, Kasenene JM. Medicinal plant diversity and uses in the Sango bay area, Southern Uganda. Journal of Ethnopharmacology. 2007;**113**(3):521-540

[3] Kumar V, Marković T, Emerald M, Dey A. Herbs: Composition and dietary importance. In: Caballero B, Finglas PM, Toldrá F, editors. Encyclopedia of Food and Health. Oxford: Academic Press; 2016. pp. 332-337

[4] Cunha AP, Ribeiro JA, Roque OR. Plantas aromáticas em Portugal: Caracterização e Utilizações. Lisboa, Portugal: Fundação Calouste Glubenkian; 2007

[5] Dixon RA. Natural products and plant disease resistance. Nature. 2005;**411**: 843-847

[6] Ríos J-L. Chapter 1 - Essential oils: What they are and how the terms are used and defined. In: Preedy VR, editor. Essential Oils in Food Preservation, Flavor and Safety. 1st ed. Academic Press; 2016. pp. 3-10

[7] de Morais LAS. Influência dos fatores abióticos na composição química dos óleos essenciais. Horticultura Brasileira. 2009;**27**(2):S4050-S4S63

[8] Lima HRP, Kaplan MAC, Cruz A. Influência dos fatores abióticos na produção e variabilidade de terpenóides em plantas. Floresta e Ambiente. 2003; **10**(2):71-77

[9] Figueiredo AC, Barroso JG, Pedro LG, Scheffer JJC. Factors affecting secondary metabolite production in plants: Volatile components and essential oils. Flavour and Fragrance Journal. 2008;**23**(4):213-226

[10] Dhifi W, Bellili S, Jazi S, Bahloul N, Mnif W. Essential oils' chemical characterization and investigation of some biological activities: A critical review. Medicines (Basel). 2016;**3**(4):1-16

[11] Baptiste Hzounda Fokou J, Michel Jazet Dongmo P, Fekam Boyom F. Essential oil's chemical composition and pharmacological properties. In: El-Shemy HA, editor. Essential Oils: Oils of Nature. IntechOpen; 2020

[12] Turek C, Stintzing FC. Stability of essential oils: A review. Comprehensive Reviews in Food Science and Food Safety. 2013;**12**(1):40-53

[13] Cragg GM, Newman DJ. Natural products: A continuing source of novel drug leads. Biochimica et Biophysica Acta. 2013;**1830**(6):3670-3695

[14] Tepe B, Daferera D, Sokmen A, Sokmen M, Polissiou M. Antimicrobial and antioxidant activities of the essential oil and various extracts of *Salvia tomentosa* Miller (Lamiaceae). Food Chemistry. 2005;**90**(3):333-340

[15] Yang SM, Feskanich D, Willett WC, Eliassen AH, Wu TY. Association between global biomarkers of oxidative stress and hip fracture in postmenopausal women: A Prospective Study. Journal of Bone and Mineral Research. 2014;**29**(12):2577-2583

[16] Lykkesfeldt J, Svendsen O. Oxidants and antioxidants in disease: Oxidative stress in farm animals. Veterinary Journal. 2007;**173**(3):502-511

[17] Nunes XP, Silva FS, JRGdS A, LAdA R, JTd L, LJQ J, et al. Biological oxidations and antioxidant activity of natural products. In: Rao DV, editor. Phytochemicals as Nutraceuticals: Global Approaches to Their Role in Nutrition and Health. Vol. 2012. Brazil: InTech; 2012. p. 278

[18] Paur I, Carlsen MH, Halvorsen BL, Blomhoff R. Antioxidants in herbs and spices: Roles in oxidative stress and redox signaling. In: IFF B, Wachtel-Galor S, editors. Herbal Medicine: Biomolecular and Clinical Aspects. 2nd ed. Boca Raton (FL): CRC Press; 2011

[19] Bakkali F, Averbeck S, Averbeck D, Idaomar M. Biological effects of essential oils: A review. Food and Chemical Toxicology. 2008;**46**(2):446-475

[20] Brenes A, Roura E. Essential oils in poultry nutrition: Main effects and modes of action. Animal Feed Science and Technology. 2010;**158**(1-2):1-14

[21] Miguel MG, Matos F, Duarte J, Venancio F, Moiteiro C, Correia AID, et al. Antioxidant capacity of the essential oils from *Lavandula luisieri*, L. *stoechas* subsp. *lusitanica*, *L. stoechas* subsp. *lusitanica* x *L. luisieri* and *L. viridis* grown in Algarve (Portugal). Journal of Essential Oil Research. 2009;**21**(4):327-336

[22] Pombal S, Rodrigues CF, Araujo JP, Rocha PM, Rodilla JM, Diez D, et al. Antibacterial and antioxidant activity of Portuguese *Lavandula luisieri* (Rozeira) Rivas-Martinez and its relation with their chemical composition. Springerplus. 2016;**5**(1):1711

[23] Arantes S, Candeias F, Lopes O, Lima M, Pereira M, Tinoco T, et al. Pharmacological and toxicological studies of essential oil of *Lavandula stoechas* subsp. *luisieri*. Planta Medicine. 2016;**82**(14):1266-1273

[24] Rufino AT, Ferreira I, Judas F, Salgueiro L, Lopes MC, Cavaleiro C, et al. Differential effects of the essential oils of *Lavandula luisieri* and *Eryngium duriaei* subsp. *juresianum* in cell models of two chronic inflammatory diseases. Pharmaceutical Biology. 2015;**53**(8):1220-1230

[25] Costa P, Goncalves S, Valentao P, Andrade PB, Almeida C, Nogueira JM, et al. Metabolic profile and biological activities of *Lavandula pedunculata* subsp. *lusitanica* (Chaytor) Franco: Studies on the essential oil and polar extracts. Food Chemistry. 2013;**141**(3):2501-2506

[26] Galego L, Almeida V, Goncalves V, Costa M, Monteiro I, Matos F, Miguel G. Antioxidant activity of the essential oils of *Thymbra capitata, Origanum vulgare, Thymus mastichina,* and *Calamintha baetica*. In: Gardner G, Craker LE, editors. XXVII International Horticultural Congress - IHC2006: International Symposium on Plants as Food and Medicine: The Utilization and Development of Horticultural Plants for Human Health Seoul, Korea ISHS Acta Hortic; 2008. pp. 325-334

[27] Arantes SM, Picarra A, Guerreiro M, Salvador C, Candeias F, Caldeira AT, et al. Toxicological and pharmacological properties of essential oils of *Calamintha nepeta*, *Origanum virens* and *Thymus mastichina* of Alentejo (Portugal). Food and Chemical Toxicology. 2019;**133**:110747

[28] Bozovic M, Ragno R. *Calamintha nepeta* (L.) Savi and its main essential oil constituent Pulegone: Biological activities and chemistry. Molecules (Basel, Switzerland). 2017;**22**(2):290-340

[29] Wang W, Wu N, Zu YG, Fu YJ. Antioxidative activity of *Rosmarinus officinalis* L. essential oil compared to its main components. Food Chemistry. 2008;**108**(3):1019-1022

[30] Gachkar L, Yadegari D, Rezael MB, Taghizadeh M, Astaneh SA, Rasooli I. Chemical and biological characteristics of *Cuminum cyminum* and *Rosmarinus officinalis* essential oils. Food Chemistry. 2007;**102**(3):898-904

[31] Galovičová L, Borotová P, Valková V, Kačániová M. Antibiofilm and antioxidant activity of *Rosmarinus officinalis* essential oil. Potravinarstvo Slovak Journal of Food Sciences. 2021;**15**:1093-1103

[32] Hussain AI, Anwar F, Chatha SA, Jabbar A, Mahboob S, Nigam PS. *Rosmarinus officinalis* essential oil: Antiproliferative, antioxidant and antibacterial activities. Brazilian Journal of Microbiology : [publication of the Brazilian Society for Microbiology]. 2010;**41**(4):1070-1078

[33] Erkan N, Ayranci G, Ayranci E. Antioxidant activities of rosemary (*Rosmarinus Officinalis* L.) extract, blackseed (*Nigella sativa* L.) essential oil, carnosic acid, rosmarinic acid and sesamol. Food Chemistry. 2008;**110**(1):76-82

[34] Rašković A, Milanović I, Pavlović N, Ćebović T, Vukmirović S, Mikov M. Antioxidant activity of rosemary (*Rosmarinus officinalis* L.) essential oil and its hepatoprotective potential. BMC Complementary and Alternative Medicine. 2014;**14**(1):225

[35] Mata AT, Proenca C, Ferreira AR, Serralheiro MLM, Nogueira JMF, Araujo MEM. Antioxidant and antiacetylcholinesterase activities of five plants used as Portuguese food spices. Food Chemistry. 2007;**103**(3):778-786

[36] Cutillas AB, Carrasco A, Martinez-Gutierrez R, Tomas V, Tudela J. *Thymus mastichina* L. essential oils from Murcia (Spain): Composition and antioxidant, antienzymatic and antimicrobial bioactivities. PLoS One. 2018;**13**(1):e0190790

[37] Cavanagh HM, Wilkinson JM. Biological activities of lavender essential oil. Phytotherapy Research : PTR. 2002;**16**(4):301-308

[38] Hajhashemi V, Ghannadi A, Sharif B. Anti-inflammatory and analgesic properties of the leaf extracts and essential oil of *Lavandula angustifolia* Mill. Journal of Ethnopharmacology. 2003;**89**(1):67-71

[39] Upson T, Andrews S. The taxonomy of the genus *Lavandula* L. In: Lis-Balchin M, editor. Lavander, the Genus *Lavandula*, Medicinal and Aromatic Plants- Pndustrial profiles. Londres and Nova Iorque: Taylor and Francis; 2002. p. 2

[40] Castroviejo S, Iberica F. Plantas vasculares de la Península Ibérica e Islas Baleares. Madrid: Real Jardín Botánico, CSIC; 2010

[41] Franco JA. Nova Flora de Portugal (Continente e Açores). Portugal: Lisboa Editora; 1984

[42] Zuzarte MR, Dinis AM, Cavaleiro C, Salgueiro LR, Canhoto JM. Trichomes, essential oils and *in vitro* propagation of *Lavandula pedunculata* (*Lamiaceae*). Industrial Crops and Products. 2010;**32**(3):580-587

[43] Nogueira JM, Romano A. Essential oils from micropropagated plants of *Lavandula viridis*. Phytochemical Analysis. 2002;**13**(1):4-7

[44] Goncalves S, Serra H, Nogueira JMF, Almeida R, Custodio L, Romano A. Headspace-SPME of *in vitro* shoot-cultures and micropropagated plants of

*Lavandula viridis*. Biologia Plantarum. 2008;**52**(1):133-136

[45] Zuzarte M, Goncalves MJ, Cavaleiro C, Canhoto J, Vale-Silva L, Silva MJ, et al. Chemical composition and antifungal activity of the essential oils of *Lavandula viridis* L'Her. Journal of Medical Microbiology. 2011;**60**(Pt 5):612-618

[46] Costa P, Grosso C, Gonçalves S, Andrade PB, Valentão P, Bernardo-Gil MG, et al. Supercritical fluid extraction and hydrodistillation for the recovery of bioactive compounds from *Lavandula viridis* L'Hér. Food Chemistry. 2012;**135**:112-121

[47] Ćavar S, Vidic D, Maksimović M. Volatile constituents, phenolic compounds, and antioxidant activity of *Calamintha glandulosa* (Req.) Bentham. Journal of the Science of Food and Agriculture. 2013;**93**(7):1758-1764

[48] Negro C, Notarnicola S, De Bellis L, Miceli A. Intraspecific variability of the essential oil of *Calamintha nepeta* subsp. *nepeta* from Southern Italy (Apulia). Natural Product Research. 2013;**27**(4-5):331-339

[49] Hammer K, Laghetti G, Pistrick K. *Calamintha nepeta* (L.) Savi and *Micromeria thymifolia* (Scop.) Fritsch cultivated in Italy. Genetic Resources and Crop Evolution. 2005;**52**(2):215-219

[50] Marongiu B, Piras A, Porcedda S, Falconieri D, Maxia A, Goncalves MJ, et al. Chemical composition and biological assays of essential oils of *Calamintha nepeta* (L.) Savi subsp. *nepeta* (Lamiaceae). Natural Product Research. 2010;**24**(18):1734-1742

[51] Araniti F, Lupini A, Sorgona A, Statti GA, Abenavoli MR. Phytotoxic activity of foliar volatiles and essential oils of *Calamintha nepeta* (L.) Savi. Natural Product Research. 2013;**27**(18):1651-1656

[52] Araniti F, Grana E, Reigosa MJ, Sanchez-Moreiras AM, Abenavoli MR. Individual and joint activity of terpenoids, isolated from *Calamintha nepeta* extract, on *Arabidopsis thaliana*. Natural Product Research. 2013;**27**(24):2297-2303

[53] Formisano C, Oliviero F, Rigano D, Saab AM, Senatore F. Chemical composition of essential oils and in vitro antioxidant properties of extracts and essential oils of *Calamintha origanifolia* and *Micromeria myrtifolia*, two *Lamiaceae* from the Lebanon flora. Industrial Crops and Products. 2014;**62**:405-411

[54] Pacifico S, Galasso S, Piccolella S, Kretschmer N, Pan S-P, Marciano S, et al. Seasonal variation in phenolic composition and antioxidant and anti-inflammatory activities of *Calamintha nepeta* (L.) Savi. Foodservice Research International. 2015;**69**(1):121-132

[55] Hernández MD, Sotomayor JA, Hernández Á, Jordán MJ. Rosemary (*Rosmarinus officinalis* L.) Oils. In: Preedy VR, editor. Essential Oils in Food Preservation, Flavor and Safety. San Diego: Academic Press; 2016. pp. 677-688

[56] Alu'datt MH, Rababah T, Alhamad MN, Gammoh S, Al-Mahasneh MA, Tranchant CC, et al. Chapter 15 - Pharmaceutical, nutraceutical and therapeutic properties of selected wild medicinal plants: Thyme, spearmint, and rosemary. In: Grumezescu AM, Holban AM, editors. Therapeutic, Probiotic, and Unconventional Foods. Academic Press; 2018. pp. 275-290

[57] Cunha AP, Ribeiro JA, Roque OR. Plantas aromáticas em portugal Caracterização e Utilização. Lisboa, Portugal: Fundação Calouste Gulbenkian; 2007

[58] Ribeiro BS, Ferreira MdF, Moreira JL, Santos L. Simultaneous distillation–Extraction of essential oils from *Rosmarinus officinalis* L. Cosmetics. 2021;**8**(4):117-127

[59] Jordan MJ, Lax V, Rota MC, Loran S, Sotomayor JA. Effect of bioclimatic area on the essential oil composition and antibacterial activity of *Rosmarinus officinalis* L. Food Control. 2013;**30**(2): 463-468

[60] Martins MR, Tinoco MT, Almeida AS, Cruz-Morais J. Chemical composition, antioxidant and antimicrobial properties of three essential oils from Portuguese flora. Journal of Phcog. 2012;**3**(1):39-44

[61] Tohidi B, Rahimmalek M, Trindade H. Review on essential oil, extracts composition, molecular and phytochemical properties of *Thymus species* in Iran. Industrial Crops and Products. 2019;**134**:89-99

[62] Stahl-Biskup E, Laakso I. Essential oil polymorphism in finnish thymus species. Planta Medica. 1990;**56**(5):464-468

[63] Sostaric I, Arsenijevic J, Acic S, Stevanovic ZD. Essential oil polymorphism of *Thymus pannonicus* All. (*Lamiaceae*) in Serbia. Journal of Essential Oil-Bearing Plants. 2012;**15**(2):237-243

[64] Salgueiro LR, Vila R, Felix T, Figueiredo AC, Barroso JG, Canigueral S, et al. Variability of essential oils of *Thymus caespititius* from Portugal. Phytochemistry. 1997;**45**(2):307-311

[65] Salgueiro LR, Vila R, Tomas X, Canigueral S, Paiva J, Proenca da Cunha A, et al. Chemotaxonomic study on *Thymus villosus* from Portugal. Biochemical Systematics and Ecology. 2000;**28**(5):471-482

[66] Salgueiro LMR. Essential oils of endemic *Thymus* species from Portugal. Flavour and Fragrance Journal. 1992;**7**(3):159-162

[67] Li X, He T, Wang X, Shen M, Yan X, Fan S, et al. Traditional uses, chemical constituents and biological activities of plants from the Genus *Thymus*. Chemistry & Biodiversity. 2019;**16**(9):e1900254

[68] FLORA-ON. Flora de Portugal Interativa. Sociedade Portuguesa de Botânica [Available from: http://www.flora-on.pt

[69] Mendez-Tovar I, Martin H, Santiago Y, Ibeas A, Herrero B, Asensio-S-Manzanera MC. Variation in morphological traits among *Thymus mastichina* (L.) L. populations. Genetic Resources and Crop Evolution. 2015;**62**(8):1257-1267

[70] Salgueiro LR, Pinto E, Gonçalves MJ, Costa I, Palmeira A, Cavaleiro C, et al. Antifungal activity of the essential oil of *Thymus capitellatus* against *Candida*, *Aspergillus* and dermatophyte strains. Flavour and Fragrance Journal. 2006;**21**(5):749-753

[71] Benchaar C, Calsamiglia S, Chaves AV, Fraser GR, Colombatto D, McAllister TA, et al. A review of plant-derived essential oils in ruminant nutrition and production. Animal Feed Science and Technology. 2008;**145**(1-4):209-228

[72] Chiang HM, Chiu HH, Lai YM, Chen CY, Chiang HL. Carbonyl species characteristics during the evaporation of essential oils. Atmospheric Environment. 2010;**44**(18):2240-2247

[73] Jori A, Bianchetti A, Prestini PE, Gerattini S. Effect of eucalyptol (1,8-cineole) on the metabolism of other drugs in rats and in man.

European Journal of Pharmacology. 1970;**9**(3):362-366

[74] Moss M, Oliver L. Plasma 1,8-cineole correlates with cognitive performance following exposure to rosemary essential oil aroma. Therapeutics Advances in Psychopharmacology. 2012;**2**(3):103-113

[75] Murata S, Shiragami R, Kosugi C, Tezuka T, Yamazaki M, Hirano A, et al. Antitumor effect of 1, 8-cineole against colon cancer. Oncology Reports. 2013;**30**(6):2647-2652

[76] Juergens UR. Anti-inflammatory properties of the monoterpene 1.8-cineole: Current evidence for co-medication in inflammatory airway diseases. Drug Research (Stuttg). 2014;**64**(12):638-646

[77] Brown SK, Garver WS, Orlando RA. 1,8-cineole: An underappreciated anti-inflammatory therapeutic. Journal of Biomolecular Research & Therapeutics. 2017;**06**(01):154-160

[78] Vuuren SFV, Viljoen AM. Antimicrobial activity of limonene enantiomers and 1, 8-cineole alone and in combination. Flavour and Fragrance Journal. 2007;**22**(6):540-544

[79] Santos FA, Rao VS. 1,8-cineol, a food flavoring agent, prevents ethanol-induced gastric injury in rats. Digestive Diseases and Sciences. 2001;**46**(2):331-337

[80] Khan A, Vaibhav K, Javed H, Tabassum R, Ahmed ME, Khan MM, et al. 1,8-cineole (eucalyptol) mitigates inflammation in amyloid Beta toxicated PC12 cells: Relevance to Alzheimer's disease. Neurochemical Research. 2014;**39**(2):344-352

[81] Cai ZM, Peng JQ, Chen Y, Tao L, Zhang YY, Fu LY, et al. 1,8-Cineole: A review of source, biological activities, and application. Journal of Asian Natural Products Research. 2021;**23**(10):938-954

[82] Bhowal M, Gopal M. Eucalyptol: Safety and pharmacological profile. Journal of Pharmaceutical Sciences. 2015;**5**:125-131

[83] Zuzarte M, Goncalves MJ, Cruz MT, Cavaleiro C, Canhoto J, Vaz S, et al. *Lavandula luisieri* essential oil as a source of antifungal drugs. Food Chemistry. 2012;**135**(3):1505-1510

[84] Gonzalez-Coloma A, Martin-Benito D, Mohamed N, Garcia-Vallejo MC, Soria AC. Antifeedant effects and chemical composition of essential oils from different populations of *Lavandula luisieri* L. Biochemical Systematics and Ecology. 2006;**34**(8):609-616

[85] Baldovini N, Lavoine-Hanneguelle S, Ferrando G, Dusart G, Lizzani-Cuvelier L. Necrodane monoterpenoids from *Lavandula luisieri*. Phytochemistry. 2005;**66**(14):1651-1655

[86] Sanz J, Soria AC, Garcia-Vallejob MC. Analysis of volatile components of *Lavandula luisieri* L. by direct thermal desorption-gas chromatography-mass spectrometry. Journal of Chromatography. A. 2004;**1024**(1-2):139-146

[87] Lavoine-Hanneguelle S, Casabianca H. New compounds from the essential oil and absolute of *Lavandula luisieri* L. Journal of Essential Oil Research. 2004;**16**(5):445-448

[88] Zuzarte M, Sousa C, Cavaleiro C, Cruz MT, Salgueiro L. The Anti-Inflammatory Response of Lavandula luisieri and Lavandula pedunculata Essential Oils. Plants. 2022;**11**(3):370-376

[89] Feijao MD, Teixeira G, Vasconcelos T, Rodrigues L, Correia AI, Sanches J, et al.

Essential oil variability and trichomes morphology from *Lavandula pedunculata* (Mill.) Cav. grown at Mata Experimental do Escaroupim (Portugal). Planta Medica. 2011;**77**(12):1297

[90] Zuzarte M, Goncalves MJ, Cavaleiro C, Dinis AM, Canhoto JM, Salgueiro LR. Chemical composition and antifungal activity of the essential oils of *Lavandula pedunculata* (Miller) Cav. Chemistry & Biodiversity. 2009;**6**(8):1283-1292

[91] Boukhobza Z, Boulenouar N, AbdelkrİM C, Kadrİ Z. Essential Oil of *Rosmarinus officinalis* L. from West Highlands of Algeria: Chemical Composition and Antifungal Activity against *Fusarium oxysporum* f. sp. *albedinis*. Natural Volatiles and Essential Oils. 2021;**8**(3):44-45

[92] Ambrico A, Trupo M, Martino M, Sharma N. Essential oil of *Calamintha nepeta* (L.) Savi subsp. *nepeta* is a potential control agent for some postharvest fruit diseases. Organic. Agriculture. 2019;**10**(1):35-48

[93] Kitic D, Stojanovic G, Palic R, Randjelovic V. Chemical composition and microbial activity of the essential oil of *Calamintha nepeta* (L.) Savi ssp *nepeta* var. *subisodonda* (Borb.) Hayek from Serbia. Journal of Essential Oil Research. 2005;**17**(6):701-703

[94] Miguel MG, Duarte F, Venancio F, Tavares R. Chemical composition of the essential oils from *Thymus mastichina* over a day period. Proceedings of the International Conference on Medicinal and Aromatic Plants Possibilities and Limitations of Medicinal and Aromatic Plant Production in the 21st Century. 2002;**576**:87-90

[95] Fraternale D, Giamperi L, Ricci D, Rocchi MBL, Guidi L, Epifano F, et al. The effect of triacontanol on micropropagation and on secretory system of *Thymus mastichina*. Plant Cell, Tissue and Organ Culture. 2003;**74**(1):87-97

[96] Machado M, Dinis AM, Santos-Rosa M, Alves V, Salgueiro L, Cavaleiro C, et al. Activity of *Thymus capitellatus* volatile extract, 1,8-cineole and borneol against *Leishmania* species. Veterinary Parasitology. 2014;**200**(1-2):39-49

[97] Delgado F. Conservação e valorização de *Asphodelus bento-rainhae* P.Silva e *Lavandula luisieri* (Rozeira) Rivas -Martínez da Beira Interior. Lisboa, Portugal: Universidade Técnica de Lisboa; 2010

[98] Garcia-Vallejo MI, Garcia-Vallejo MC, Sanz J, Bernas M, Velasco– Negueruela A. Necrodane(1,2,2,3,4-pentamethylcyclopentana) derivaties in *Lavandula luisieri*, new compounds to the plant kingdom. Phytochemistry. 1994;**36**:43-45

[99] Negueruela AV, Perez-Alonso MJ, Maria J, Rico MM. Essential oil of Iberian *Lamiaceae* with pulegone as basic component. Anales de Bromatologia. 1987;**39**:357-372

[100] Akgül A, Pooter HL, Buyck LF. The essential oils of *Calamintha nepeta* subsp. *nepeta* and ssp. *glandulosa* and *Ziziphora clinopodioides* from Turkey. Journal of Essential Oil Research. 1991;**3**:7-10

[101] Risstorcelli D, Tomi F, Casanova J. Essential oils of *Calamintha nepeta* subsp. *nepeta* and subsp. *glandulosa* from Corsica (France). Journal of Essential Oil Research. 1996;**8**:363-366

[102] Flamini G, Cioni PL, Puleio R, Morelli I, Panizzi L. Antimicrobial activity of the essential oil of *Calamintha nepeta*

and its constituent pulegone against bacteria and fungi. Phytotherapy Research: PTR. 1999;**13**(4):349-351

[103] Cozzolino F, Fellous R, Vernin G, Parkanyi C. GC/MS Analysis of the Volatile Constituents of *Calamintha nepeta* (L.) Savi ssp. *nepeta* from Southeastern France. Journal of Essential Oil Research. 2000;**12**(4):481-486

[104] Couladis M, Tzakou O. Essential oil of *Calamintha nepeta* subsp *glandulosa* from Greece. Journal of Essential Oil Research. 2001;**13**(1):11-12

[105] Kitic D, Jovanovic T, Ristic M, Palic R, Stojanovic G. Chemical composition and antimicrobial activity of the essential oil of *Calamintha nepeta* (L.) Savi ssp. *glandulosa* (Req.) P.W. Ball from Montenegro. Journal of Essential Oil Research. 2002;**14**(2):150-152

[106] Fraternale D, Giamperi L, Ricci D, Manunta A. Composition of the essential oil as a taxonomic marker for *Calamintha nepeta* (L.) Savi ssp. *nepeta*. Journal of Essential Oil Research. 1998;**10**:568-570

[107] Arantes SM, Picarra A, Guerreiro M, Salvador C, Candeias F, Caldeira AT, Martins MR. Toxicological and pharmacological properties of essential oils of *Calamintha nepeta, Origanum virens* and *Thymus mastichina* of Alentejo (Portugal). Food and Chemical Toxicology. 2019;**133**:110747-110755

[108] Arantes S, Picarra A, Candeias F, Caldeira AT, Martins MR, Teixeira D. Antioxidant activity and cholinesterase inhibition studies of four flavouring herbs from Alentejo. Natural Product Research. 2017;**31**(18):2183-2187

[109] Macedo-Arantes S, Picarra A, Caldeira AT, Candeias AE, Martins MR. Essential oils of Portuguese flavouring plants: Potential as green biocides in cultural heritage. European Physical Journal Plus. 2021;**136**(11):1-15

[110] Galovičová L, Borotová P, Valková V, Vukovic NL, Vukic M, Štefániková J, et al. *Thymus vulgaris* essential oil and its biological activity. Plants (Basel, Switzerland). 2021;**10**(9):1959-1976

[111] Salgueiro LR, Vila R, Tomas X, Canigueral S, DaCunha AP, Adzet T. Composition and variability of the essential oils of *Thymus* species from Section *Mastichina* from Portugal. Biochemical Systematics and Ecology. 1997;**25**(7):659-672

[112] Moldão-Martins M, Beirão-da-Costa S, Neves C, Cavaleiro C, Lg S. Luísa Beirão-da-Costa M. Olive oil flavoured by the essential oils of *Mentha × piperita* and *Thymus mastichina* L. Food Quality and Preference. 2004;**15**(5):447-452

[113] Chiu HH, Chiang HM, Lo CC, Cheri CY, Chiang HL. Constituents of volatile organic compounds of evaporating essential oil. Atmospheric Environment. 2009;**43**(36):5743-5749

[114] Ghasemzadeh A, Ghasemzadeh N. Flavonoids and phenolic acids: Role and biochemical activity in plants and human. Journal of Medicinal Plant Research. 2011;**5**(31):6697-6703

[115] Pelicano H, Carney D, Huang P. ROS stress in cancer cells and therapeutic implications. Drug Resistance Updates. 2004;7(2):97-110

[116] Moyo M, Ndhlala AR, Finnie JF, Van Staden J. Phenolic composition, antioxidant and acetylcholinesterase inhibitory activities of *Sclerocarya birrea* and *Harpephyllum caffrum* (*Anacardiaceae*) extracts. Food Chemistry. 2010;**123**(1):69-76

[117] Albano SM, Miguel MG. Biological activities of extracts of plants grown in Portugal. Industrial Crops and Products. 2011;**33**(2):338-343

[118] Gonzalez-Coloma A, Delgado F, Rodilla JM, Silva L, Sanz J, Burillo J. Chemical and biological profiles of *Lavandula luisieri* essential oils from western Iberia Peninsula populations. Biochemical Systematics and Ecology. 2011;**39**(1):1-8

[119] Miguel MG, Falcato-Simoes M, Figueiredo AC, Barroso JMG, Pedro LG, Carvalho LM. Evaluation of the antioxidant activity of *Thymbra capitata*, *Thymus mastichina* and *Thymus camphoratus* essential oils. Journal of Food Lipids. 2005;**12**(3):181-197

[120] Smeriglio A, Alloisio S, Raimondo FM, Denaro M, Xiao J, Cornara L, et al. Essential oil of *Citrus lumia* Risso: Phytochemical profile, antioxidant properties and activity on the central nervous system. Food and Chemical Toxicology. 2018;**119**:407-416

[121] Ruberto G, Baratta MT. Antioxidant activity of selected essential oil components in two lipid model systems. Food Chemistry. 2000;**69**(2):167-174

[122] Shahat AA, Ibrahim AY, Hendawy SF, Omer EA, Hammouda FM, Abdel-Rahman FH, et al. Chemical composition, antimicrobial and antioxidant activities of essential oils from organically cultivated fennel cultivars. Molecules (Basel, Switzerland). 2011;**16**(2):1366-1377

[123] Lobo AP, García YD, Sánchez JM, Madrera RR, Valles BS. Phenolic and antioxidant composition of cider. Journal of Food Composition and Analysis. 2009;**22**:644-648

[124] Podsedek A. Natural antioxidants and antioxidant capacity of *Brassica* vegetables: A review. LWT- Food Science and Technology. 2007;**40**(1):1-11

[125] Raut JS, Karuppayil SM. A status review on the medicinal properties of essential oils. Industrial Crops and Products. 2014;**62**:250-264

[126] Miguel MG. Antioxidant and anti-inflammatory activities of essential oils: A short review. Molecules (Basel, Switzerland). 2010;**15**(12):9252-9287

[127] Dohi S, Terasaki M, Makino M. Acetylcholinesterase inhibitory activity and chemical composition of commercial essential oils. Journal of Agricultural and Food Chemistry. 2009;**57**(10):4313-4318

[128] Li CJ, Chang JK, Wang GJ, Ho ML. Constitutively expressed COX-2 in osteoblasts positively regulates Akt signal transduction via suppression of PTEN activity. Bone. 2011;**48**(2):286-297

[129] Utar Z, Majid MI, Adenan MI, Jamil MF, Lan TM. Mitragynine inhibits the COX-2 mRNA expression and prostaglandin E(2) production induced by lipopolysaccharide in RAW264.7 macrophage cells. Journal of Ethnopharmacology. 2011;**136**(1): 75-82

[130] Calvello R, Panaro MA, Carbone ML, Cianciulli A, Perrone MG, Vitale P, et al. Novel selective COX-1 inhibitors suppress neuroinflammatory mediators in LPS-stimulated N13 microglial cells. Pharmacological Research. 2012;**65**(1):137-148

[131] Jeon J, Park KA, Lee H, Shin S, Zhang T, Won M, et al. Water extract of *Cynanchi atrati* Radix regulates inflammation and apoptotic cell death through suppression of IKK-mediated NF-kB signaling. Journal of Ethnopharmacology. 2011;**137**:626-634

[132] Rios JL. Effects of triterpenes on the immune system. Journal of Ethnopharmacology. 2010;**128**(1):1-14

[133] Kumar N, Drabu S, Mondal SC. NSAID's and selectively COX-2 inhibitors as potential chemoprotective agents against cancer. Arabian Journal of Chemistry. 2013;**6**(1):1-23

[134] Suresh V, Sruthi V, Padmaja B, Asha VV. *In vitro* anti-inflammatory and anti-cancer activities of *Cuscuta reflexa* Roxb. Journal of Ethnopharmacology. 2011;**134**(3):872-877

[135] Homnan N, Thongpraditchote S, Chomnawang M, Thirapanmethee K. *In vitro* Anti-inflammatory effects of Thai herb essential oils. Asiaweek. 2020;**47**(2):153-163

[136] Valacchi G, Virgili F, Cervellati C, Pecorelli A. OxInflammation: From subclinical condition to pathological biomarker. Frontiers in Physiology. 2018;**9**:858

[137] Rohleder N. Stress and inflammation: The need to address the gap in the transition between acute and chronic stress effects. Psychoneuroendocrinology. 2019;**105**:164-171

[138] Monti D, Ostan R, Borelli V, Castellani G, Franceschi C. Inflammaging and human longevity in the omics era. Mechanisms of Ageing and Development. 2017;**165**(Pt B):129-138

[139] Franceschi C, Campisi J. Chronic inflammation (inflammaging) and its potential contribution to age-associated diseases. The Journals of Gerontology. Series A, Biological Sciences and Medical Sciences. 2014;**69**(Suppl 1):S4-S9

[140] Fulop T, Witkowski JM, Olivieri F, Larbi A. The integration of inflammaging in age-related diseases. Seminars in Immunology. 2018;**40**:17-35

[141] Baylis D, Bartlett DB, Patel HP, Roberts HC. Understanding how we age: Insights into inflammaging. Longevity & Healthspan. 2013;**2**(8):1-8

[142] de Lavor EM, Fernandes AWC, de Andrade Teles RB, Leal A, de Oliveira Junior RG, Gama ESM, et al. Essential oils and their major compounds in the treatment of chronic inflammation: A review of antioxidant potential in preclinical studies and molecular mechanisms. Oxidative Medicine and Cellular Longevity. 2018;**2018**:6468593

[143] de Sousa DP. Analgesic-like activity of essential oils constituents. Molecules (Basel, Switzerland). 2011;**16**(3):2233-2252

[144] Schabauer L, Steflitsch W, Buchbauer G. Essential oils and compounds against pains in animal studies. Natural Product Communications. 2017;**12**(7):1137-1143

[145] Sarmento-Neto JF, do Nascimento LG, Felipe CF, de Sousa DP. Analgesic potential of essential oils. Molecules (Basel, Switzerland). 2015;**21**(1):E20

[146] da Silveira ESR, Lima TC, Nobrega FR, Brito AEM, Sousa DP. Analgesic-like activity of essential oil constituents: An update. International Journal of Molecular Sciences. 2017; **18**(12):1-40

[147] Lenardão EJ, Savegnago L, Jacob RG, Victoria FN, Martinez DM. Antinociceptive effect of essential oils and their constituents: An update review. Journal of the Brazilian Chemical Society. 2015;**27**(3):435-474

[148] Santos FA, Rao VS. Antiinflammatory and antinociceptive effects of 1,8-cineole a terpenoid oxide present in many plant

essential oils. Phytotherapy Research : PTR. 2000;**14**(4):240-244

[149] Peana AT, D'Aquila PS, Panin F, Serra G, Pippia P, Moretti MD. Anti-inflammatory activity of linalool and linalyl acetate constituents of essential oils. Phytomedicine. 2002;**9**(8):721-726

[150] Miller SG, Pritts TA. Linalool: A novel natural anti-inflammatory agent?: Commentary on "Anti-inflammatory effects of linalool in RAW 264.7 macrophages and lipopolysaccharide-induced lung injury model". The Journal of Surgical Research. 2013;**185**(1):e69-e70

[151] da Silveira R, Andrade LN, de Sousa DP. A review on anti-inflammatory activity of monoterpenes. Molecules (Basel, Switzerland). 2013;**18**(1):1227-1254

[152] Guimaraes AG, Quintans JS, Quintans LJ Jr. Monoterpenes with analgesic activity: A systematic review. Phytotherapy Research : PTR. 2013;**27**(1):1-15

[153] Loizzo MR, Tundis R, Menichini F, Saab AM, Statti GA, Menichini F. Cytotoxic activity of essential oils from *Labiatae* and *Lauraceae* families against *in vitro* human tumor models. Anticancer Research. 2007;**27**(5A):3293-3299

[154] Dragomanova S, Tancheva L, Georgieva M, Klisurov R. Analgesic and anti-inflammatory activity of monoterpenoid myrtenal in rodents. Journal of Imab. 2019;**25**(1):2406-2413

[155] Pergolizzi JV, Taylor R, JA LQ, Raffa RB, Group NR. The role and mechanism of action of menthol in topical analgesic products. Journal of Clinical Pharmacy and Therapeutics. 2018;**43**(3):313-319

[156] Sengupta R, Sheorey SD, Hinge M. Analgesic and anti-inflammatory plants: An updated review. International Journal of Pharmaceutical Sciences Review and Research. 2012;**12**:114-119

[157] Padhan DK. Topical analgesic activity of essential oil extracted from *Spharenthus Indicus* (Asteraceae). Asian Journal of Pharmaceutical and Clinical Research. 2017;**10**(5):275-277

[158] Cardia GFE, Silva-Filho SE, Silva EL, Uchida NS, Cavalcante HAO, Cassarotti LL, et al. Effect of Lavender (*Lavandula angustifolia*) essential oil on acute inflammatory response. Evidence-based Complementary and Alternative Medicine. 2018;**2018**:1413940

[159] Juhas S, BujnaKova D, Rehak P, Cikos S, Czikkova S, Vesela J, et al. Anti-inflammatory effects of thyme essential oil in mice. Acta Veterinaria Brno. 2008;**77**(3):327-334

[160] Fachini-Queiroz FC, Kummer R, Estevao-Silva CF, Carvalho MD, Cunha JM, Grespan R, et al. Effects of thymol and carvacrol, constituents of *Thymus vulgaris* L. essential oil, on the inflammatory response. Evidence-based Complementary and Alternative Medicine. 2012;**2012**:657026

[161] Goncalves JC, de Meneses DA, de Vasconcelos AP, Piauilino CA, Almeida FR, Napoli EM, et al. Essential oil composition and antinociceptive activity of *Thymus capitatus*. Pharmaceutical Biology. 2017;**55**(1):782-786

[162] Juergens LJ, Worth H, Juergens UR. New perspectives for mucolytic, anti-inflammatory and adjunctive therapy with 1,8-cineole in COPD and asthma: Review on the new therapeutic approach. Advances in Therapy. 2020;**37**(5):1737-1753

Chapter 12

# Antioxidant Effect and Medicinal Properties of Allspice Essential Oil

*Yasvet Yareni Andrade Avila, Julián Cruz-Olivares and César Pérez-Alonso*

## Abstract

*Pimenta dioica L. Merrill.* Myrtaceae family, known for its berries called pimenta or allspice, is one of the oldest spices in the world, widely used for its culinary and medicinal qualities. The main commercial product obtained from this spice is its essential oil, the reason for the interest in essential oil is based on the versatility of its use in different industrial areas (food, cosmetics, perfumery, and pharmaceuticals) due to its harmless and beneficial effects for health. In addition, it contains compounds that have shown broad biological activity, which turns out to be useful in the treatment of diseases related to the excessive formation of oxygen radicals. As a result, the extraction process and operating conditions have a significant impact on the bioactivity of these molecules. As a consequence, selecting the correct mix of variables to improve oil extraction and functionality is essential. The most of study on this essential oil is being focused on resolving these issues, as well as purification and identification. This chapter will cover the methods for obtaining *P. dioica* essential oil, as well as the chemical profile of the oil and its biological properties, which include its effects on humans, plants, animals, insects, and microorganisms.

**Keywords:** *Pimenta dioica*, essential oil, eugenol, antioxidant effect, chemical composition

## 1. Introduction

Allspice (*Pimenta dioica* L. Merrill or Pepper officinalis) belongs to the Myrtaceae family native to the West Indies and Central America [1]. In Mexico, it is found in the wild and is cultivated toward the east and southeast [2]. The commercial spice, known in Mexico as pimienta gorda and in English as "allspice," is a small tree that grows up to 6–12 m tall [3] with small, whitish flowers with a peculiar aroma; its dry, almost spherical, reddish-brown berries are the commercial pepper spice, known in Mexico as pimienta gorda; and in English as "allspice" for flavors that resemble a mixture of cinnamon, cloves, and nutmeg [4]. This spice is known for its antioxidant qualities, which are attributed to the presence of bioactive components, most especially polyphenolic compounds [5]. *P. dioica* is one of the most important spices as a source of essential oils high in eugenol, a phenolic compound having antibacterial and antioxidant properties against a variety of pathogens. *P. dioica* produced in Central

America is sent to the international markets because its use in the local market is minimal. Its manufacture and drying, on the other hand, are entirely traditional [6].

Allspice contains its oils both in its leaves and in the berry itself [7], with fairly variable returns (1.5–4.5%) [8]. According to reports, the oil content varies depending on where it originates located [9]. González and Pino [10] and Shaik et al. [11] also discovered that environmental parameters, harvesting procedures, drying, and the age of the trees all influence the chemical composition of the oil.

It is important to mention that the oil obtained from the leaf is a brownish-yellow liquid with a dry, woody, warm, and spicy aromatic smell, while the oil extracted from the berry is yellow in color with a warm spicy-sweet smell and a note of sweet and fresh output, and placed in the spicy-sweet and warm group [12].

Allspice essential oil is utilized in the food sector, specifically in the meat and tanner industries, and also in perfumery and cosmetic products [13]. In addition, it has been useful for the treatment of gastrointestinal disorders, cramps, flatulence, indigestion, and nausea. Likewise, it has managed to help in cases of depression, nervous exhaustion, tension, neuralgia, and stress, it is also used as a natural repellent [14]. Anesthetic, analgesic, antibacterial, antioxidant, antiseptic, acaricide, carminative, muscle relaxant, rubefacient, stimulant, and tonic are some of the medicinal effects of this essential oil [15].

The versatility of essential oils' use in different industrial areas (pharmaceuticals, food, and cosmetics) has sparked interest in recent years, not only because of the possibility of obtaining aromatic compounds, but also because of their use as antioxidants, food preservatives, and medicines, as well as their use as crop and plant protectants, incorporating them into the packaging material of the products [16].

## 2. Essential oil extraction

Steam distillation, hydrodistillation, and the use of organic solvents are the most common extraction procedures. To produce the essential oil, steam distillation uses saturated steam at atmospheric pressure. When the steam breaks the cells of the plant walls, the water generates steam, and the essence is freed, the extraction is complete [17, 18]. They allow the process to be favorable for the creation of alcohols and acids when the esters disintegrate by employing high temperatures and the presence of water, resulting in a decrease in the extraction of the oil, which is one of the limitations of distillation by steam entrainment [19].

In recent years, several novel techniques for extracting essential oils have been developed, including ultrasound-assisted extraction, microwave-assisted extraction, and extraction using supercritical fluids, with the goal of reducing extraction time, reducing solvent consumption, increasing extraction yield, and improving the quality of the extracts [20]. Traditional organic solvent extraction, while easy, has drawbacks, such as expensive prices, is not environmentally friendly, and is nonselective, requiring post-treatment processes for product purification. Nonrecyclable organic solvent disposal can also be hazardous to human health and the environment.

On the other hand, at the laboratory and pilot scale, supercritical fluid extraction of flavonoid compounds presents a viable alternative for a more efficient and environmentally friendly extraction process. The volatile concentrate obtained from allspice by supercritical fluids was compared to the oil obtained by the hydrodistillation method by Marongiu [21], with the primary differences being the amount of eugenol, 77.9% against 45.4%. It was also demonstrated that by employing supercritical $CO_2$, the extract has an additional benefit in that it is free of hydrocarbons, which can conceal or degrade the oil's natural aroma.

Other studies compared the effects of microwave energy supply and hydrodistillation radiation time (MHD) on the performance and composition of allspice essential oil [22]. While there were no significant differences in the yields (2.68% versus 3.25%) and chemical composition of essential oils obtained by HD and MHD, the advantage was obtained in the reduction of the extraction cost in terms of time and energy.

## 3. Allspice essential oil chemical profile

Polyphenols, lignins, and terpenoids are the most prevalent components found in allspice essential oil currently [23]. The basic component of the oil is eugenol, finding that the oil content obtained from the leaves (65–96%) is somewhat higher than that of the berry oil [14]. **Table 1** shows the chemical composition of the essential

| Country Of Origin | Component of the plant | Year | Method of extraction | Main constituents (%area) | References |
|---|---|---|---|---|---|
| Antilles | Leaves | 2007 | Commercial | Eugenol (47.78%)<br>Myrcene (26.76%) | [15] |
| Australia | Leaves | 2005 | SCD | Eugenol (77.9%)<br>β-caryophyllene (5.1%) | [21] |
| | Leaves | 2005 | HD | Eugenol (45.4%)<br>β-caryophyllene (8.9%) | [21] |
| Brazil | Fruit | 2011 | HD | Eugenol (76.98%)<br>β-pinene (6.52%)<br>5-indanol(5.88%)<br>limonene (4.09%) | [24] |
| | Leaves | 2014 | HD | Eugenol (60.8%)<br>Myrcene (19.3%)<br>limonene (6.48%) | [25] |
| | Fruit | 2020 | HD | Eugenol (76.88%)<br>β-Pinene (6.52%) | [26] |
| China | Fruit | 2013 | HD | Eugenol (28.84%)<br>Methyl eugenol (43.01%) | [22] |
| Cuba | Leaves | 1997 | HD | Eugenol (28.04%)<br>1,8-cineole (14.5%)<br>α-humulene(10.12%)<br>γ-cadinene (11.12%) | [27] |
| | Leaves | 1997 | SCD | Eugenol (93.87%)<br>thymol (1.82%) | [27] |
| | Leaves | 1997 | SE | Eugenol (91.68%)<br>thymol (2.72%) | [27] |
| | Leaves | 2003 | HD | Eugenol (34.14%)<br>1,8-cineole (14.69%)<br>α-humulene (10.12%) | [28] |
| Guatemala | Leaves | 2020 | HD | Eugenol (71.4%)<br>Myrcene (10.0%) | [29] |
| | Fruit | 2020 | HD | Eugenol (65.9%)<br>Myrcene (10.1%) | [29] |

| Country Of Origin | Component of the plant | Year | Method of extraction | Main constituents (%area) | References |
|---|---|---|---|---|---|
| India | Fruit | 2013 | HD | Eugenol (68.4%)<br>chavicol (10.4%)<br>methyl eugenol (6.1%) | [30] |
| | Fruit | 2015 | Commercial | Eugenol (35.42%)<br>methyl eugenol (28.02%)<br>β-caryophyllene (8.66%)<br>β-Myrcene (8.55%) | [31] |
| Jamaica | Leaves | 1991 | SD | Eugenol (66.38%–79.24%) | [32] |
| | Leaves | 2007 | Commercial | Eugenol (76.02%)<br>methyl eugenol (7.14%)<br>β-caryophyllene (6.47%) | [33] |
| | Leaves | 2007 | HD | Eugenol (79.81–83.68) | [34] |
| | Berries | 2007 | Commercial | Eugenol (86.44%)<br>β-caryophyllene (7.70%)<br>Methyl eugenol (3.87%) | [35] |
| | Leaves | 2009 | Commercial | Eugenol (76.0%) | [36] |
| | Berries | 2016 | SCD | Eugenol (63.94%)<br>β-caryophyllene (4.65%) | [37] |
| | Berries | 2016 | HD | Eugenol (66.8%)<br>β-caryophyllene (4.69%) | [37] |
| México | Berries | 1997 | SD | Methyl eugenol (48.3%)<br>Myrcene (17.7%)<br>Eugenol (17.3%) | [38] |
| | Berries | 1997 | HD | Methyl eugenol (62.7%)<br>Myrcene (16.5%)<br>eugenol (8.3%) | [38] |
| | Berries | 1997 | SCD | Methyl eugenol (67.9%)<br>Eugenol (14.9%)<br>Myrcene (6.0%) | [38] |
| | Berries | 2011 | SD | Methyl eugenol (62.7%)<br>Eugenol (8.3%) | [39] |
| | Fruit | 2011 | HD | Methyl eugenol (48.7%)<br>Myrcene (17.1%)<br>Eugenol (16.3%) | [40] |
| | Leaves | 2013 | HD | Eugenol (94.86%)<br>α-terpineol (2.45%) | [41] |
| | Berries | 2018 | HD | Methyl eugenol (65.14%)<br>β-Myrcene (12.72%) | [42] |
| | Fruit | 2020 | HD | Eugenol (48.5%)<br>Methyl eugenol (35.0%) | [43] |
| Sri Lanka | Leaves | 2015 | HD | Eugenol (85.33%)<br>β-caryophylene (4.36%)<br>Cineole (4.19%) | [44] |
| USA | Leaves | 2012 | HD | Eugenol (62.1%)<br>Methyl eugenol (22.9%) | [45] |

*SD = steam distillation; HD = hydrodistillation; SCD = supercritical carbon dioxide; SE = solvent extraction.*

**Table 1.**
*Chemical composition of the essential oil of* Pimenta dioica.

oil of *P. dioica* obtained by using gas chromatography coupled to mass spectrometry (GC-MS) analysis technique, as well as data from the literature obtained from various researchers denoting the main compounds present in the essential oil, according to the extraction method, geographical origin, and plant part used in the extraction. Essential oils are complicated combinations with a high number of elements, and their physicochemical qualities are controlled by factors, such as harvest time, soil type, and fruit storage conditions and time [24]. The quality of Jamaican berries is greater than that of other islands, and they are preferred for commerce. Allspice's oil content and flavor deteriorate when it is stored for an extended period of time [1].

Because of the extraction process used, the quantity and quality of compounds found vary. Essential oil composition has an important role in determining the spice's pharmacological potential [16]. The essential oil of *P. dioica* extracted using HD, SCD, SE, and SD have significant qualitative and quantitative changes in their chemical composition. Hydrodistillation was the most used procedure. Eugenol, methyl eugenol, and myrcene are the three main constituents of this oil.

## 4. Antioxidant effect

Spices and herbs are recognized as sources of natural antioxidants [46]. Some of the biological functions of essential oils are dependent on their antioxidant properties. These properties are attributable to some essential oil components' inherent potential to prevent or delay aerobic oxidation of organic matter. However, it is important to be cautious before thinking that essential oils' antioxidant properties are just a result of their chemical components. However, taking into account its composition can help to estimate its antioxidant capacity [47].

In terms of free radical scavenging activity against the radicals DPPH, ABTS, and superoxide anion, the composition and antioxidant activity of the essential oil obtained by hydrodistillation of the berries were studied [48]. A total of 45 components were discovered. Eugenol (74.71, 73.35%) was the most common component found, followed by methyl eugenol (4.08, 9.54%) and caryophyllene (4.08, 9.54%). The antioxidant evaluation revealed that the oil had a high rate of radical scavenging. The total phenolic content, total reducing power, and metal chelating capacity were also calculated, and the metal chelating capabilities and reducing power were both found to be extremely high. The essential oil has a substantial antioxidant activity that is comparable to pure eugenol, according to the results.

Another study showed a positive correlation between the anticancer and antioxidant effects of allspice essential oil [42]. As a member of the Myrtaceae family, this oil has been shown to have a great cytotoxic effect against cancer cells. As a result, it might be considered a natural source of anticancer medicines. According to research, consuming foods containing synthetic antioxidants can result in health problems, such as cancer owing to the accumulation of free radicals in the body. As a result, research has been done to return to using natural compounds as an alternative for synthetic substances and as a source of novel food preservatives. These essential oils with high inhibitory percentages can now be utilized to replace synthetic additives since they help to eliminate pollutants and chemical residues, which can cause issues and diseases [17].

Allspice is a powerful hydroxyl radical scavenger. The berries of *P. dioica* had a high level of antioxidant activity and scavenging activity for 1,1-diphenyl-2-picryl-hydrazyl (DPPH) radical [49]. The capacity of *P. dioica* leaf essential oil to combat

DPPH (2,2-diphenyl-1-picrylhydrazyl), hydroxyl (OH), and superoxide radicals was studied to determine its antioxidant characteristics [33]. The intrinsic characteristics of many of their bioactive components, particularly phenols, to block or delay oxidation, are responsible for the antioxidant potential of *P. dioca* essential oil.

Although not all phenolic molecules had antibacterial activity, antioxidant activity was significantly related to total phenol content. *P. dioica* leaf extracts include phenolic chemicals that can be employed as antioxidants in the food, cosmetics, and pharmaceutical industries [50].

Allspice essential oil showed a high concentration of antioxidants The antioxidant characteristics of the essential oil were compared to those of propyl gallate, a synthetic antioxidant, and it was discovered that the essential oil's free radical scavenging activity was dependent on the concentration and higher than that of propyl gallate [51]. Antioxidants were found in abundance in allspice essential oil. (i.e. > 75 mmol/100 g) [52]. Applications in medicine have been reported due to the presence of antioxidant chemicals in *P. dioica*'s essential oil.

## 5. Medicinal properties

The essential oil of allspice is a significant source of phytochemicals in medicine. Phytochemicals are a large group of plant-derived bioactive that may have disease-fighting properties [53]. Plants are one of the most important natural sources of secondary metabolites for medical purposes, due to their biological capacity to combat lethal or endemic diseases, as well as disorders that impact living beings.

Anticancer, antidermatophytic, antihemorrhagic, anti-inflammatory, antimicrobial, antimutagenic, antipyretic, central nervous system depressant, hypoglycemic, hypotensive, an inhibitor of the enzyme histone acetyltransferase, and inhibitor of the enzyme histidine have all been discovered as pharmacological effects of allspice essential oil [54–57].

### 5.1 Nematicidal activity

In other studies, Park et al. [35] discovered allspice essential oil looks to be effective as a natural nematicide for B. xylophilus, but more research on systemic action, phytotoxicity, and formulation is needed to improve nematicidal potency and stability while reducing cost.

### 5.2 Antimicrobial activity

The presence of antioxidant properties and antimicrobial effects of allspice suggests that it can be used against human pathogenic bacteria and for the control of other diseases and the support of immunity for rejuvenation. The ability of allspice to alleviate bacterial infections and its use in traditional medicine in different parts of the world was observed. Due to its use, it is possible that this plant has anti-QS properties [58]. Its important bacteriostatic and inhibitory properties of pathogenic and decomposition microorganisms against *Bacillus subtilis*, *Clostridium botulinum*, *Escherichia coli*, *Listeria monocytogenes*, *Salmonella typhimurium*, and *Staphylococcus aureus* were also reported [59].

In another study, the essential oil extracted from *P. dioica* (Myrtaceae) was evaluated for its antimicrobial activities using a panel of gram-positive pathogens, gram-negative strains, and fungi [60]. Antimicrobial activity was measured by the minimum inhibitory concentration required to inhibit the growth of microorganisms. The cytotoxicity of the essential oil was tested *ex vivo* using the THP-1 macrophage cell model. The results showed that it had antimicrobial activity.

Allspice oil reduced xanthine oxidase activity, resulting in a decrease in superoxide radical formation. Both the synthesis of conjugated dienes and the development of secondary products from lipid peroxidation were effectively inhibited by allspice oil. Infections caused by *Klebsiella*, *Pseudomonas*, *A. niger*, *A. flavus*, and *T. versicolor* can be treated with *P. offcinalis* as an alternative to synthetic medications, according to the literature, depending on the chemical composition of the allspice oil [61]. Allspice has been shown to suppress *Escherichia coli*, *Salmonella enterica*, and *Listeria monocytogenes* [62].

The antibacterial activity of allspice essential oil was tested by using the agar diffusion method against three microorganism strains. *B. cereus*, *S. typhimurium*, and *S. aureus* were found to be inhibited by it. *B. cereus* was found to be the microbe most vulnerable to the presence of oil in the microdilution. The predominant component of *P. dioica* was eugenol, which had an abundance proportion of 94.86% as determined by GC-MS [41].

### 5.3 Anticancer activity

Cancer is a worldwide health issue. In breast (MCF-7), hepatocellular (HepG-2), colon (HCT-116), prostate (PC-3), and cervical cancer cell lines, allspice essential oil was examined for cytotoxicity. The MTT assay was used on HeLa cells. The essential oil had cytotoxic action against the cell lines that were examined [42]. The results showed that the essential oil of Mexican allspice has cytotoxic activity ($IC_{50} < 15$ μg/mL) against the cancer cell lines examined.

### 5.4 Antifungal activity

The antifungal efficacy of *P. dioica* leaf essential oil against toxin-producing Aspergillus flavus was investigated in one study. Antifungal activity of *P. dioica* leaf EO was shown on *A. flavus in vitro* experiments (IISRaf1). These tests revealed that this EO could be used as a food additive because of its antifungal properties and capacity to decrease ergosterol formation, which would extend the storage life of post-harvest items [63].

Allspice oil was found to have a superior antifungal impact against *Fusarium oxysporum*, *Fusarium verticillioides*, *Penicillium expansum*, *Penicillium brevicompactum*, *Aspergillus flavus*, and *Aspergillus fumigatus*. As a result, its efficacy is comparable to that of synthetic fungicides often used to treat severe human mycoses. The MIC values of *P. dioica*, which were detected against all pathogens tested, are very remarkable [64].

The fungal activity and chemical composition of the essential oil obtained from the fruits of *P. dioica* in the mycelial development of *Fusarium oxysporum* f. sp. *lycopersici*, *Fusarium oxysporum* f. sp. *passiflorae*, *Fusarium subglutinans* f. sp. *ananas*, *Fusarium oxysporum* f. sp. *vasinfectum*. The oil contained 76.88% eugenol and suppressed fungal mycelial development by up to 97.78% in an average of 7.2 days, according to the findings. As a result, the oil could be used as a natural fungicide [26].

*Aspergillus niger*, *Candida blanki*, *Candida tropicalis*, *Candida cylindracea*, *Saccharomyces cerevisiae*, and *Candida albicans* were found to have strong inhibitory activity, while *Candida glabrata*, *Candida krusei*, *C. albicans*, and *C. albicans* were found to have moderate inhibitory activity. With an activity index of 1.20–2.80, all of the test fungi were suppressed. This suggests that ketoconazole has a stronger antifungal effect against *C. albicans*, *Candida glabrata*, *C. tropicalis*, *Candida cylindracea*, *C. albicans*, and *Aspergillus niger* [65].

It has also recently become a research hub for the development of novel insecticides for ecologically friendly plants. Its insecticidal action has been demonstrated in numerous studies, and it can be utilized as a natural repellant [66].

### 5.5 Antidiabetic effect

Allspice berry extract was reported to inhibit protein glycation, indicating its potential to be used as an effective antidiabetic agent [67]. Studies have shown that individual flavonoids inhibit glycation by 50%.

### 5.6 Acaricidal effect

The essential oil derived from *P. dioica* berries was found to be highly harmful to *R. microplus* 10-day-old larvae in this investigation. As a result, the findings point to a viable new technique that could be utilized as an alternative to synthetic acaricides for tick management. The main components, methyl eugenol (62.7%) and eugenol (62.7%), could be responsible for acaricidal activity (8.3%) [39].

The active components of allspice essential oil were used in one investigation to cause mortality and limit the development of *B. microplus* to a level comparable to commercial acaricides. The phenylpropanoid molecules responsible for this activity, eugenol and methyl eugenol, could be studied for use as Acarina chemosterilants and as templates for the synthesis of further acaricides. All extracts, commercial acaricides, and methyl eugenol were found to be less effective in suppressing oviposition and causing tick mortality than berry essential oil. Eugenol, a component contained in more than 65% of the oil composition, is responsible for the effectiveness of berry essential oil [68].

## 6. Conclusions

Over the years, researchers have studied the enormous range of biological activities of allspice essential oil and its potential applications. *P. dioica* essential oil contains a large number of medicinal compounds. Currently, the need to extract compounds of interest from plant materials drives the continuous search for economically and ecologically viable extraction technologies. We have given a quick rundown of the medicinal characteristics of allspice essential oil, with a focus on the chemical components that have biological activity.

## Author details

Yasvet Yareni Andrade Avila*, Julián Cruz-Olivares and César Pérez-Alonso
Facultad de Química, Universidad Autónoma del Estado de México,
Toluca, Estado de México, Mexico

*Address all correspondence to: yasvety@gmail.com

## References


[1] Krishnamoorthy B, Rema J. Allspice. In: Handbook of Herbs and Spices. United Kingdom: Woodhead Publishing. Sawston; 1981. pp. 5-15

[2] Barco MJM. La Pimienta De Jamaica [*Pimenta Dioica* (L.) Merrill, Myrtaceae] En La Sierra Norte De Puebla (México). Anales Del Jardin Botanico de Madrid. 1998;**56**(2):337-349

[3] Badura M. Pimenta officinalis Lindl. (pimento, myrtle pepper) from early modern latrines in Gdańsk (northern Poland). Vegetation History and Archaeobotany. 2003;**12**(4):249-252. DOI: 10.1007/s00334-003-0023-6

[4] Lim TK. *Pimenta dioica*. Edible Medicinal and Non Medicinal Plants. 2012;**3**:655-664. DOI: 10.1007/978-94-007-2534-8_89

[5] Mercado-Mercado G, Carrillo L d l R, Wall-Medrano A, Díaz JAL, Álvarez-Parrilla E. Compuestos polifenólicos y capacidad antioxidante de especias típicas consumidas en México. Nutricion Hospitalaria. 2013;**28**(1):36-46. DOI: 10.3305/nh.2013.28.1.6298

[6] Jarquín-Enríquez L, Ibarra-Torres P, Jiménez-Islas H, Flores-Martínez NL. *Pimenta dioica*: A review on its composition, phytochemistry, and applications in food technology. International Food Research Journal. 2021;**28**(5):893-904

[7] Cruz-Olivares J, Perez AC, Barrera-Pichardo JF. Extracción de aceite esencial de hojas y fruto de pimienta gorda mexicana *Pimenta dioica* L. Revista Ciencias Agrícolas: Merrill; 2007

[8] Weyerstahl P, Marschall-Weyerstahl H, Christiansen C, Oguntimein BO, Adeoye AO. Volatile constituents of Eugenia uniflora leaf oil. Planta Medica. 1988;**54**(6):546-549. DOI: 10.1055/s-2006-962544

[9] Green, Espinosa CLyF. Jamaican and central American pimento (allspice; Pimenta dioica): Characterization of flavour differences and other distinguishing features, development. Food Science. 1988;**1**:3-20

[10] González, Pino AyJA. Efecto del envejecimiento del árbol de Pimenta dioica L. sobre el rendimiento y composición del aceite esencial de la fruta, Revista de Agroquímica y Tecnología de Alimentos. 1990;**30**:277-281

[11] Shaikh J, Bhosale R, Singhal R. Microencapsulation of black pepper oleoresin. Food Chemistry. 2006;**94**:105-110

[12] Rema J, Krishnamoorthy B. Allspice. In: Handbook of Herbs and Spices. 2nd ed. Vol. 2. Sawston, United Kingdom: Woodhead Publishing; 2012. pp. 166-192. DOI: 10.1533/9780857095688.166

[13] Rao PS, Navinchandra S, Jayaveera KN. HPTLC analysis of the essential oil from *Pimenta dioica* leaf. Journal of Applied Pharmaceutical Science. 2012;**2**(12):141-148. DOI: 10.7324/JAPS.2012.21226

[14] Rao PS, Navinchandra S, Jayaveera K. An important spice, *Pimenta dioica* (Linn.) Merill: A Review. International Current Pharmaceutical Journal. 2012a;**1**(8):221-225. DOI: 10.3329/icpj.v1i8.11255

[15] Oussalah M, Caillet S, Saucier L, Lacroix M. Antimicrobial effects of

selected plant essential oils on the growth of a *Pseudomonas putida* strain isolated from meat. Meat Science. 2006;**73**(2):236-244. DOI: 10.1016/j.meatsci.2005.11.019

[16] Contreras-Moreno BZ. Chemical Composition of Essential Oil of Genus Pimenta (Myrtaceae): Review. Potential of Essential Oils. London, UK: IntechOpen Limited; 2018. DOI: 10.5772/intechopen.78004

[17] Minango MB. Aplicación de los aceites esenciales de albahaca (Ocimum Basilicum), cilandro (Coriandrum Sativum), y guayaba (Psidium Guajava) como antioxidante en salchichas de pollo tipo Frankfurt. Quito: Universidad de las Amércas; 2017

[18] Segovia SRA. Utilización de aceites esenciales naturales como conservantes en la elaboración de salchichas de pollo. Universidad Politecnica Salesiana Ecuador. 2014;**2014**:129. Available from: http://dspace.ups.edu.ec/bitstream/123456789/7009/1/UPS-CT003676.pdf

[19] Quispe Díaz KF, Taco Huamani RJ. Evaluación del tiempo de extracción, factor de empaquetamiento, humedad del rizoma, en el rendimiento de la extracción del aceite esencial de jengibre (Zingiber Officinale Roscoe) por arrastre de vapor, caracterización fisicoquimica del aceite esencial y aplicación antioxidante en el aceite de oliva. Tesis de Licenciatura, Universidad Nacional de San Agustín de Arequipa. Repositorio Institucional-Universidad Nacional de San Agustín de Arequipa. 2017

[20] Velasco RJ, Villada HS, Carrera JE. Aplicaciones de los fluidos supercríticos en la agroindustria. Informacion Tecnologica. 2007;**18**(1):53-66. DOI: 10.4067/s0718-07642007000100009

[21] Marongiu B, Piras A, Porcedda S, Casu R, Pierucci P. Comparative analysis of supercritical $CO_2$ extract and oil of *Pimenta dioica* leaves. Journal of Essential Oil Research. 2005;**17**(5):530-532. DOI: 10.1080/10412905.2005.9698985

[22] Jiang ZT, Feng X, Li R, Wang Y. Composition comparison of essential oils extracted by classical hydro distillation and microwave-assisted Hydrodistillation from *Pimenta dioica*. Journal of Essential Oil-Bearing Plants. 2013;**16**(1):45-50. DOI: 10.1080/0972060X.2013.764178

[23] Zhang L, Lokeshwar L, B. Medicinal properties of the Jamaican pepper plant *Pimenta dioica* and allspice. Current Drug Targets. 2012;**13**(14):1900-1906. DOI: 10.2174/138945012804545641

[24] Monteiro OS, Souza AG, Soledade LEB, Queiroz N, Souza AL, Mouchrek Filho VE, et al. Chemical evaluation and thermal analysis of the essential oil from the fruits of the vegetable species *Pimenta dioica* Lindl. Journal of Thermal Analysis and Calorimetry. 2011;**106**(2):595-600. DOI: 10.1007/s10973-011-1438-4

[25] Faria LR, Machado RUD, Pimenta PH, Oliveira LDAR, de Castro Peixoto J, de Paula JER, et al. Structural organization and phytochemical analysis of *Pimenta dioica* L Merrill Myrtaceae leaves collected from Gois State, Brazil. Journal of Medicinal Plants Research. 2014;**8**(38):1134-1147. DOI: 10.5897/jmpr2014.5395

[26] Barros Gomes PR, Da Silva Barros Junior FR, Reis JB, Oliveira Everton G, Santos de Oliveira RW, Costa Louzeiro H, et al. Chemical composition and biological activity of the essential oil of the fruits *Pimenta dioica* against formae speciales of fungus Fusarium

oxysporum. Revista Colombiana de Ciencias Químico-Farmacéuticas. 2020;**49**(1):89-100. DOI: 10.15446/rcciquifa.v49n1.87010

[27] Pino JA, Garcia J, Martinez MA. Solvent extraction and supercritical carbon dioxide extraction of *Pimenta dioica* merrill, Leaf. Journal of Essential Oil Research. 1997;**9**(6):689-691. DOI: 10.1080/10412905.1997.9700812

[28] Hernández L, Rodríguez M, García D, Pino J. Actividad antidermatofítica in vitro de aceites esenciales. Revista Cubana de Plantas Medicinales. 2003;**8**(2):1059865. ISSN: 1028-4796. Available from: http://scielo.sld.cu/scielo.php?script=sci_arttext&pid=S1028-47962003000200004&lng=es&nrm=iso

[29] Mérida-Reyes MS, Muñoz-Wug MA, Oliva-Hernández BE, Gaitán-Fernández IC, Simas DLR, Ribeiro da Silva AJ, et al. Composition and Antibacterial Activity of the Essential Oil from *Pimenta dioica* (L.) Merr. from Guatemala. Medicines. 2020;**7**(10):59. DOI: 10.3390/medicines7100059

[30] Amma KPP, Rani MP, Sasidharan I, Sreekumar MM. Comparative chemical composition and in vitro antioxidant activities of essential oil isolated from the leaves of *Cinnamomum tamala* and *Pimenta dioica*. Natural Product Research. 2013;**27**(3):290-294. DOI: 10.1080/14786419.2012.668691

[31] Misharina TA, Alinkina ES, Medvedeva IB. Antiradical properties of essential oils and extracts from clove bud and pimento. Applied Biochemistry and Microbiology. 2015;**51**(1):119-124. DOI: 10.1134/S0003683815010093

[32] Tucker AO, Maciarello MJ, Landrum LR. Volatile leaf oils of Caribbean Myrtaceae. II. *Pimenta dioica* (L.) Merr. of Jamaica. Journal of Essential Oil Research. 1991;**3**(3):195-196. DOI: 10.1080/10412905.1991.9700504

[33] Jirovetz L, Buchbauer G, Stoilova I, Krastanov A, Stoyanova A, Schmidt E. Spice plants: Chemical composition and antioxidant properties of Pimenta Lindl. Essential oils, part 2: *Pimenta racemosa* (Mill.) JW Moore leaf oil from Jamaica. Nutrition-Vienna. International Journal of Biosciences (IJB). 2007;**31**((7/8)):293

[34] Minott DA, Brown HA. Differentiation of fruiting and non-fruiting *Pimenta dioica* (l.) merr. Trees based on composition of leaf volatiles. Journal of Essential Oil Research. 2007;**19**(4):354-357. DOI: 10.1080/10412905.2007.9699303

[35] Park IK, Kim J, Lee SG, Shin SC. Nematicidal activity of plant essential oils and components from Ajowan (Trachyspermum ammi), Allspice (Pimenta dioica) and Litsea (Litsea cubeba) essential oils against pine wood nematode (Bursaphelenchus Xylophilus). Journal of Nematology. 2007;**39**(3):275-279

[36] Höferl M, Buchbauer G, Jirovetz L, Schmidt E, Stoyanova A, Denkova Z, et al. Correlation of antimicrobial activities of various essential oils and their main aromatic volatile constituents. Journal of Essential Oil Research. 2009;**21**(5):459-463. DOI: 10.1080/10412905.2009.9700218

[37] Toni-Moy AS, Lowe H, Watson C. Quantification and characterization of *Pimenta dioica* (allspice) essential oil extracted via hydrodistillation, solvent and super critical fluid extraction methodologies. American Journal of Essential Oils and Natural Products. 2016;**4**(3):27-30

[38] García-Fajardo J, Martínez-Sosa M, Estarrón-Espinosa M, Vilarem G, Gaset A, de Santos JM. Comparative study of the oil and supercritical $CO_2$ extract of Mexican pimento (*Pimenta dioica* Merrill). Journal of Essential Oil Research. 1997;**9**(2):181-185. DOI: 10.1080/10412905.1997.9699456

[39] Martinez-Velazquez M, Castillo-Herrera GA, Rosario-Cruz R, Flores-Fernandez JM, Lopez-Ramirez J, Hernandez-Gutierrez R, et al. Acaricidal effect and chemical composition of essential oils extracted from *Cuminum cyminum*, *Pimenta dioica* and *Ocimum basilicum* against the cattle tick Rhipicephalus (Boophilus) microplus (Acari: Ixodidae). Parasitology Research. 2011;**108**(2):481-487. DOI: 10.1007/s00436-010-2069-6

[40] Sánchez-Sáenz EO, Pérez-Alonso C, Cruz-Olivares J, Román-Guerrero A, Baéz-González JG, Rodríguez-Huezo ME. Establishing the most suitable storage conditions for microencapsulated allspice essential oil entrapped in blended biopolymers matrices. Drying Technology. 2011;**29**(8):863-872. DOI: 10.1080/07373937.2010.545495

[41] Vazquez-Cahuich DA, Moreno JE, Hidalgo DC, Martinez JRV, Borges-Argaez R, Farfan MC. Antimicrobial activity and chemical composition of the essential oils of malvaviscus arboreus cav, *Pimenta dioica* (L.) merr., *Byrsonima crassifolia* (L.) kunth and *Psidium guajava* l. Tropical and Subtropical Agroecosystems. 2013;**16**(3):505-513

[42] Morsy NFS, Hammad KSM. Volatile constituents, radical scavenging and cytotoxic activities of Mexican allspice (*Pimenta dioica* L. Merrill) berries essential oil. Journal of Essential Oil-Bearing Plants. 2018;**21**(4):859-868. DOI: 10.1080/0972060X.2018.1524793

[43] Milenković A, Stanojević J, Stojanović-Radić Z, Pejčić M, Cvetković D, Zvezdanović J, et al. Chemical composition, antioxidative and antimicrobial activity of allspice (*Pimenta dioica* (L.) Merr.) essential oil and extract. Advanced Technologies. 2020;**9**(1):27-36. DOI: 10.5937/savteh2001027m

[44] Dharmadasa RM, Abeysinghe DC, Dissanayake DMN, Abeywardhane KW, Fernando NS. Leaf essential oil composition, antioxidant activity, total phenolic content and total flavonoid content of *Pimenta dioica* (L.) Merr (Myrtaceae): A superior quality spice grown in Sri Lanka. Universal Journal of Agricultural Research. 2015;**3**(2):49-52. DOI: 10.13189/ujar.2015.030203

[45] Kloucek P, Smid J, Frankova A, Kokoska L, Valterova I, Pavela R. Fast screening method for assessment of antimicrobial activity of essential oils in vapor phase. Food Research International. 2012;**47**(2):161-165. DOI: 10.1016/j.foodres.2011.04.044

[46] Mariutti LRB, Barreto GPDM, Bragagnolo N, Mercadante AZ. Free radical scavenging activity of ethanolic extracts from herbs and spices commercialized in Brazil. Brazilian Archives of Biology and Technology. 2008;**51**(6):1225-1232. DOI: 10.1590/S1516-89132008000600018

[47] Amorati R, Foti MC, Valgimigli L. Antioxidant activity of essential oils. Journal of Agricultural and Food Chemistry. 2013;**61**(46):10835-10847. DOI: 10.1021/jf403496k

[48] Padmakumari KP, Sasidharan I, Sreekumar MM. Composition and antioxidant activity of essential oil of pimento (*Pimenta dioica* (L) Merr.) from Jamaica. Natural Product

Research. 2011;**25**(2):152-160. DOI: 10.1080/14786419.2010.526606

[49] Miyajima Y, Kikuzaki H, Hisamoto M, Nakatani N. Antioxidative polyphenols from berries of *Pimenta dioica*. BioFactors. 2004;**22**(1-4):301-303. DOI: 10.1002/biof.5520220159

[50] Sánchez-Zarate A, Hernández-Gallegos MA, Carrera-Lanestosa A, López-Martínez S, Chay-Canul AJ, Esparza-Rivera JR, et al. Antioxidant and antibacterial activity of aqueous, ethanolic and acetonic extracts of *Pimenta dioica* L. leaves. International Food Research Journal. 2020;**27**(5):825-834

[51] Feng X, Jiang Z, Li R, Wang X. Antioxidant activities and free radical scavenging potentials of natural spice. essential oil, China Condiment. Journal of Agricultural and Food Chemistry. 2010;**4**:52-55

[52] Dragland S, Senoo H, Wake K, Holte K, Blomhoff R. Several culinary and medicinal herbs are important sources of dietary antioxidants. The Journal of Nutrition. 2003;**133**(5):1286-1290

[53] Murali VS, Devi VNM, Parvathy P, Murugan M. Phytochemical screening, FTIR spectral analysis, antioxidant and antibacterial activity of leaf extract of *Pimenta dioica* Linn. Materials Today: Proceedings. 2021;**45**:2166-2170. DOI: 10.1016/j.matpr.2020.10.038

[54] Hari Kumar B, Badarudin A, Jose A. DPPH radical scavenging activity and antibacterial activity of *Pimenta dioica* (L.) merr. Oriental Journal of Chemistry. 2010;**26**(4):1501-1505

[55] Mueller M, Hobiger S, Jungbauer A. Anti-inflammatory activity of extracts from fruits, herbs and spices. Food Chemistry. 2010;**122**(4):987-996. DOI: 10.1016/j.foodchem.2010.03.041

[56] Shamaladevi N, Lyn DA, Shaaban KA, Zhang L, Villate S, Rohr J, et al. Ericifolin: A novel antitumor compound from allspice that silences androgen receptor in prostate cancer. Carcinogenesis. 2013;**34**(8):1822-1832. DOI: 10.1093/carcin/bgt123

[57] Kikuzaki H, Miyajima Y, Nakatani N. Phenolic glycosides from berries of *Pimenta dioica*. Journal of Natural Products. 2008;**71**(5):861-865. DOI: 10.1021/np0705615

[58] Vasavi HS, Arun AB, Rekha PD. Inhibition of quorum sensing in Chromobacterium violaceum by *Syzygium cumini* L. and *Pimenta dioica* L. Asian Pacific Journal of Tropical Biomedicine. 2013;**3**(12):954-959. DOI: 10.1016/S2221-1691(13)60185-9

[59] Tajkarimi MM, Ibrahim SA, Cliver DO. Antimicrobial herb and spice compounds in food. Food Control. 2010;**21**(9):1199-1218. DOI: 10.1016/j.foodcont.2010.02.003

[60] Lorenzo-Leal AC, Palou E, López-Malo A, Bach H. Antimicrobial, cytotoxic, and anti-inflammatory activities of *Pimenta dioica* and *Rosmarinus officinalis* essential oils. BioMed Research International. 2019;**2019**(8):1-8. DOI: 10.1155/2019/1639726

[61] Vadlapudi V, Kaladhar DSVGK. Phytochemical evaluation and molecular characterization of some important medicinal plants. Asian Pacific Journal of Tropical Disease. 2012;**2**(1):S26-S32. DOI: 10.1016/S2222-1808(12)60118-2

[62] Friedman M, Henika PR, Mandrell RE. Bactericidal activities of plant essential oils and some of their isolated constituents against

*Campylobacter jejuni*, *Escherichia coli*, *Listeria monocytogenes*, and *Salmonella enterica*. Journal of Food Protection. 2002;**65**(10):1545-1560. DOI: 10.4315/0362-028X-65.10.1545

[63] Sarathambal C, Rajagopal S, Viswanathan R. Mechanism of antioxidant and antifungal properties of *Pimenta dioica* (L.) leaf essential oil on *Aspergillus flavus*. Journal of Food Science and Technology. 2021;**58**(7):2497-2506. DOI: 10.1007/s13197-020-04756-0

[64] Zabka M, Pavela R, Slezakova L. Antifungal effect of *Pimenta dioica* essential oil against dangerous pathogenic and toxinogenic fungi. Industrial Crops and Products. 2009;**30**(2):250-253. DOI: 10.1016/j.indcrop.2009.04.002

[65] Kamble VA, Patil SD. Spice-derived essential oils: Effective antifungal and possible therapeutic agents. Journal of Herbs, Spices and Medicinal Plants. 2008;**14**(3-4):129-143. DOI: 10.1080/10496470802598677

[66] Pavela R. Insecticidal and repellent activity of selected essential oils against of the pollen beetle, *Meligethes aeneus* (Fabricius) adults. Industrial Crops and Products. 2011;**34**(1):888-892. DOI: 10.1016/j.indcrop.2011.02.014

[67] Dearlove RP, Greenspan P, Hartle DK, Swanson RB, Hargrove JL. Inhibition of protein glycation by extracts of culinary herbs and spices. Journal of Medicinal Food. 2008;**11**(2):275-281. DOI: 10.1089/jmf.2007.536

[68] Brown HA, Minorr DA, Ingram CW, Williams LAD. Biological activities of the extracts and constituents of pimento, *Pimenta dioica* l. against the southern cattle tick, boophilus microplus. Insect Science and its Application. 1998;**18**(1):9-16. DOI: 10.1017/s1742758400007402